COMPUTER SCIENCE, TECHNOLOGY AND APPLICATIONS

HUMAN-COMPUTER INTERACTION

COMPUTER SCIENCE, TECHNOLOGY AND APPLICATIONS

Additional books and e-books.com in this series can be found on Nova's website under the Series tab.

COMPUTER SCIENCE, TECHNOLOGY AND APPLICATIONS

HUMAN-COMPUTER INTERACTION

TURGUT OZSEVEN
EDITOR

nova
science publishers
New York

Library of Congress Cataloging-in-Publication Data

ISBN: 978-1-53616-495-4

Published by Nova Science Publishers, Inc. † New York

CONTENTS

PREFACE

Human Computer Interaction is used in all areas of our daily lives as a result of the rapid development of technology and computer systems. Human Computer Interaction is an interdisciplinary field of study involving the design and implementation of interactive technologies. The field of Human Computer Interaction is related to many areas such as human behavior, psychology, cognitive sciences, computer technologies, software engineering, ergonomics, graphic / industrial design, sociology and educational sciences. Researchers of this subject both observe the interaction of people with computers and design different technologies and examine the interaction of people with these technologies. The Human Computer Interaction system has four main components: user, task, tool, context.

Human Computer Interaction aims to develop interactive technologies through design, evaluation and implementation processes. The development of interactive technologies depends on usability. Usability can be determined by evaluating effectiveness, efficiency and satisfaction together. Effectiveness includes how much users can accomplish the tasks they are expected to do using the application; efficiency, how long the user has done the job; Satisfaction refers to the measure of the user's ideas when using the application.

One of the major shortcomings in HCI is the transformation of theoretical knowledge into practice. The purpose of the book is to introduce students, teachers, researchers, and practitioners to new advances in HCI. The book includes theoretical and practical studies prepared with the academic contributions of scientists working in different fields. It was decided to publish each chapter in the book after being examined by the scientific board. As an editor, my duty is to ensure breadth, while the chapter authors treat the delegated chapters with depth.

The book is designed for practitioners or researchers of all levels of expertise from novice to expert. Each of the book's individual topics could be considered as a compact, self-contained mini-book right under its title. The approach is to provide a framework and a set of techniques for evaluating and improving HCI. It presents a specific set of solutions, mostly obtained from real world projects and experimental studies, for routine applications. It further highlights promising emerging techniques for research and exploration opportunities.

The development team of this book wanted to thank their colleagues who made contributions to this book by providing continuous encouragements and thorough reviews of the chapters of the book.

Dr. Turgut Ozseven
turgutozseven@gmail.com

In: Human-Computer Interaction
Editor: Turgut Ozseven
ISBN: 978-1-53616-495-4

Chapter 1

A REVIEW ON EEG CONTROLLED BCI: DEEP LEARNING APPROACH

***Caglar Uyulan*[1]*, Turker Tekin Erguzel*[2,*]**
***and Nevzat Tarhan*[3]**
[1]Department of Mechatronics Engineering,
Bulent Ecevit University, Zonguldak, Turkey
[2]Department of Software Engineering,
Uskudar University, İstanbul, Turkey
[3]Department of Psychology, Uskudar University, İstanbul, Turkey

ABSTRACT

Advanced signal processing methodologies are required to correctly interpret and analyse the characteristics of Electroencephalography (EEG) signal. The development of EEG-based Brain-Computer Interface (BCI) algorithms, which are characterised as subject-independent, robust and accurate in classifying biomarkers remains an open issue in the literature. Deep Learning (DL) has a promising potential because of learning robust feature representation from raw data. The DL-EEG framework could be helpful through a suitable DL architecture and a data preprocessing technique to handle the EEG decoding problem in BCI

* Corresponding Author's Email: turker.erguzel@uskudar.edu.tr.

applications. The findings after review demonstrate that the utilisation of DL combined EEG framework in BCI has excellent potential and capacity to solve the classification problem of multichannel data and provides significant superiorities as compared to the conventional machine learning algorithms. This review covers information and practices on the application of DL to EEG signals to inspire researchers, who study on the BCI design.

Keywords: EEG, deep learning, neural networks, classification, BCI

INTRODUCTION

The postsynaptic electrical potential fields originated from the pyramidal neurons in the cortical areas are measured through EEG. EEG signals capture mental events emerging at millisecond timescales with a high temporal resolution. On the contrary, they exhibit a low spatial resolution due to the distance among the electric field sources and sensors. Due to the close distance between electrode type sensors, these sensors are spatially correlated. The signal interference leads to the source localisation problem, and therefore, the brain sources should be reconstructed from EEG data (Michel and He 2012).

The electrodes attached to the scalp receive the brainwaves emerged from neuronal activity as an input signal, and after that, these signals are transformed by the help of translation algorithms into the commands produced to control the output device. A Brain-Computer Interface (BCI) is constituted from four fundamental stages categorised as data acquisition, signal processing, signal classification/translation, controlling the physical-based device, protocol operation. The BCI applications show variety in a broad spectrum such as electric wheelchair control (Iturrate, Antelis, and Minguez 2009; Philips et al. 2007; Velasco-Álvarez et al. 2018; Long et al. 2012), robot grasping control (W. Zhang et al. 2017), gait rehabilitation systems (Castermans et al. 2014), paretic limb control via neural bypasses (Xingang, Lipeng, and Dan 2014; Moreno-Valenzuela and Salinas-Avila 2011; Inoue et al. 2008; Zecca et al. 2017), humanoid robot navigation and control (Chae, Jeong, and Jo 2012; Dasgupta et al. 2010), visual speller

(Eom et al. 2013; Mirghasemi, Fazel-Rezai, and Shamsollahi 2006), neuro-cognitive rehabilitation (Alkawadri 2019; Machado et al. 2010; Schröer et al. 2015; Ramesh, Krishna, and Nakirekanti 2014), rehabilitation robotic system (Valbuena et al. 2007; Meyer et al. 2012; Costa et al. 2014; Kim et al. 2015; Hortal et al. 2015), emotion recognition (Pan et al. 2018), thought-translation device (Birbaumer et al. 2003). Cloud-assisted network enabling technology for many human-centred application domains such as health-care, fitness, ergonomics, security, and sociality can cooperate with BCI (Fortino et al. 2014; Ullah et al. 2014).

BCI should represent fast reaction, high accuracy, generalisation capability, and application-based flexibility and adaptation. There exist many factors, such as the amount of training, concentration level in the data acquisition phase, type of application, translation algorithms, etc. determining the performance of a BCI. BCI performance is also affected by some limitations related to the EEG signal processing. First, EEG has a low Signal-to-Noise Ratio (SNR) due to the "artefacts", which are originated from multiple sources of activity-dependent noise concerning environmental, physiological roots. The filtering and noise reduction methods are utilised to attenuate the effects of these sources and reveal the effective brain signal collected from electrodes (Treder and Nolte 2018; Daly et al. 2013). The non-stationarity concept, defined as changing the statistical properties of a signal across time, is inherently embedded into the nature of the EEG signal (Cole and Voytek 2018; Ku and Kawasumi 2007).

A challenging problem in BCI technology is to translate the various brain activities corresponding to the mental tasks into output command signals, which drive the physical device appropriately. This problem has been defined as a feature extraction combined with pattern recognition and tried to be solved by algorithms originated from machine learning. The EEG signals contain critical information and sophisticated properties, so a feature extraction can elicit this implicit information that allows different mental tasks to be distinguished. A classifier trained based on the limited amount of subject data is not sufficiently generalising the results obtained from the data, which belongs to the same individual acquired at a different time if the factors as mentioned earlier are not taken care. The physiological varieties among individual subjects also restrict the classification performance of the BCI model that is aiming to be generalised across subjects. Novel adaptive methods are sought to cope with the inter-subject variability (Muehl et al. 2014; Tan and Nijholt 2010).

Feature extraction is operated in the time domain, frequency domain, time-frequency domain, spatial or spatiotemporal domain, respectively. After the feature extraction procedure, the reduced and valuable EEG data are fed to the classifier. Different brain activity patterns are automatically estimated and translated into control commands via classification algorithms. As a result, a BCI system performance is affected by the problem-specific effectiveness of the feature extraction and classification methods.

Through DL, it is possible to perform classification without the need for preprocessing, feature extraction, feature selection modules and to improve the BCI for achieving better generalisation abilities and more versatile. DL architectures proved their effectiveness in processing a vast amount of data in text, audio, or images format (Iba 2018; Muhammed, Ahmed, and Khalid 2017). Hierarchical representations of input data are learned by applying non-linear transformations, whose parameters are adjusted directly by minimising a cost function through DL. Deep Neural Networks (DNNs) are build from stacked layers of artificial neurons (LeCun, Bengio, and Hinton 2015). Assume that a multivariate EEG data time series are constituted from m-channel and n-sampled window length. The input of a neural network is a multidimensional array $S_i \in R^{(m \times n)}$ including the n samples matched with a window for all channels. S_i array fed into a DL model without passing through a preprocessing and feature extraction (time-frequency conversion). Various type of layers such as Fully-connected (FC), Convolutional Neural Networks (CNNs), and Recurrent Neural Networks (RNNs), etc. are utilized in the supervised classification model (Das, Maiorana, and Campisi 2017; Y. Yang et al. 2018). In FC layers, each neuron is connected to the activations of every single neuron of the previous layer as input.

Convolutional layers are structured as neurons in a particular layer combined with a pattern imposing a subset of the activations of the previous layer (sparse interactions, also referred as sparse connectivity or sparse weights). This is achieved by building the convolution kernel smaller than the input. This operation leads to reduce the memory requirements of the model and enhance the statistical efficiency. Through convolutional building blocks, the invariant representations of the spatially rearranged EEG data can be learned. In a deep CNN, units localised in the deeper layers interact indirectly with a larger part of the input data. In addition, pooling layers replace the output of the network at a specific position with a summary statistic of the neighbourhood outputs.

A recurrent layer collects current activations of the previous layers and its self-activations from a past time step. RNNs extract the temporal structure of the data (Sharma, Liu, and Yang 2018). In addition, unsupervised deep learning strategies have been built to train models without labelling procedure. Autoencoders (AEs) learn by replicating their inputs under constraints. AEs are utilised in dimensionality reduction or feature learning Deep generative models such as Restricted Boltzmann Machines (RBMs), Deep Belief Networks (DBNs), Deep Boltzmann Machines (DBMs), etc. are based on probabilistic models defined over probabilistic distribution function or probability density function of the data. Generative Adversarial Networks (GANs) works with the game theory-based scenario. Adversarial Networks (GANs) works with the game theory-based scenario. While a generator tries to produce samples from an unknown distribution, a discriminator attempts to dichotomise samples from the training data and samples from a generator. The agents compete with each other to reach their objectives. The training phase is modelled as; two-player zero-sum minimax game. After equilibrium point is accessed, the probability distribution related to the generator converges to the real data distribution (Goodfellow, Bengio, and Courville 2016; Shen et al. 2017).

This review is dedicated to investigating the properties and evaluation of DL oriented classification algorithms applied in EEG-based BCI. The other motivation is to guide in selecting the most effective DL classifier algorithm for a subject-centred BCI experiment by comparing the algorithms given in the literature and assessing their performances regarding the objective. This review also serves a detailed methodological knowledge about the different parts of a DL-EEG implementation.

The review is comprised as follows: The mechanism of the EEG-BCI system is explained in section 2, The deep learning-based classification algorithms used in BCI applications are sought in section 3 and, section 4 discusses their performance on the context.

DESCRIBING THE EEG-BASED BCI SYSTEM

The BCI system is developed for direct and real-time interaction between the brain and the external world (objects). Electrodes utilising a compact, high SNR, low-impedance current-voltage configuration have been developed and implemented to measure and analyse brain electrical activity (slow cortical potential, sensorimotor rhythm (Thomas, Fruitet, and Clerc 2012), event-related potentials, neuronal action potential, visual evoked potential (Trejo, Rosipal, and Matthews 2006; Cong et al. 2013), event-related synchronization&de-synchronization, etc.) of subjects (Gao et al. 2014; Portelli and Nasuto 2017; Schalk et al. 2004). BCI system can be divided into sub-components such as data acquisition, pre-processing, translator and operator interface. The characteristics of BCI are classified as, interoperability, scalability, real-time data flow, closed-loop analysis, adaptability, etc. (Guger, Ramoser, and Pfurtscheller 2000; Schalk and Leuthardt 2011).

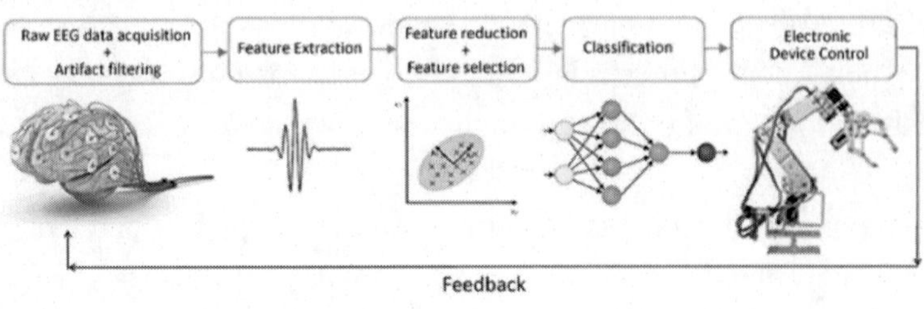

Figure 1. The flowchart of a typical EEG-based BCI.

A typical EEG-BCI whose scheme is depicted in Figure 1 is based on the pattern recognition system exploited from feature extraction, and classification algorithms related to machine learning aims to transform brain activity signals into an electronic control command. These algorithms recognise brain activity patterns after trained with the feature extracted valuable dataset. The trained classifier model is tested with a dataset different from the training dataset. If the classification algorithm cannot be adapted according to the variety of signals that the subject can produce, the static structure of the classification algorithm restricts the learning of the hypothesis (Z. Zhang et al. 2012). Information transfer rate, error rate, autonomy, and cognitive load are the main challenges encountered in the BCI design. Information transfer rate depends on how much sensor is used and the nature of the sensors (bandwidth, resolution, sampling time, etc.). The error rate is also defined as a classification error, which is related to the classification accuracy of the mental task transformed to command signal of the external device. Cognitive load refers to the complexity of the BCI system, which means that how many degrees of freedom is controlled by the user.

Feature Extraction

The most crucial stage is to decide what features are the best fit for discriminating task-based time variation of EEG dynamics for a given BCI application (Moore 2003). The features (biomarkers) such as amplitude values, band powers, power spectral density values, autoregressive and adaptive autoregressive parameters, time-frequency transformations, inverse model-based features, etc. have been used to design BCI. High dimensional feature vectors are often extracted from several channels and from several time segments and concatenated into a single vector. The other ways to extract features are combining classifications done at various time segments or building a temporal sequence of feature vectors (dynamic classifier). Mental tasks performed to control BCI depend on the particular time variations or specific frequency band of EEG signals (Gope,

Kehtarnavaz, and Nair 2005; Toderean and Chiuchisan 2017). In the segmentation of EEG signals, methods such as decomposition into equal minimal section lengths, characterisation of each section with a specific set of features (spectral estimations, auto-regressive coefficients, etc.), multivariate statistical procedures are also applied. Each stationary segment is defined in terms of a particular period and typological characteristics. Adaptive partitioning separates EEG recordings into variable-length semi-stationary sections (Kaplan and Shishkin 2000). These operations are done with parametric or non-parametric methods. Parametric methods define the semi-stationary structure of the signals, and it is useful when the process model is available (Sakkalis et al. 2008; Shiman et al. 2012) Autoregressive methods and Kalman filters take place in this category (Tzallas, Oikonomou, and Fotiadis 2006). The feature extraction methods include time-domain statistical methods (mean, standard deviation, maximum peak value, skewness, kurtosis, cross-correlation, etc.), frequency-domain methods (auto-regressive coefficients, power spectral density, band power, Fourier transform, etc.), time-frequency methods, which describe how the spectrum of power changes over time (short-time Fourier transform, Wavelet transform (S. Yan et al. 2012), etc.), spatial filtering methods, which focuses on activities occurred in the specific areas of the brain through the help of the multiple-channel electrodes (common spatial patterns (T. Yan, Jingtian, and Andong 2008), Laplacian technique, etc.) (Coyle, Prasad, and McGinnity 2005; Khushaba et al. 2010; Kline 2012), entropy (C. Yang et al. 2012) and fractal dimensions.

Feature Reduction and Feature Selection

When the feature vector is obtained, it is necessary to reduce the dimensionality to minimise the computational burden and optimise the classification performance. There are two basic strategies for feature reduction. The first one is the feature projection, which constitutes a new set of features, to determine the best combinations for representing the original feature vector. The Principal Component Analysis (PCA) is implemented at this stage. It applies an orthogonal transformation to convert the observation set of possible correlation variables to a set of linearly uncorrelated variables (Artoni, Delorme, and Makeig 2018; Bugli and Lambert 2007). PCA produces sets of features that correspond to the eigenvalues of the data covariance matrix (ter Braack, de Jonge, and van Putten 2013). The second one is the Independent Component Analysis (ICA), which determines the independent components of the signal source. This method finds the direct representation of non-Gaussian data, which is statistically independent (Caesarendra 2017; Sun, Liu, and Beadle 2005). EEG data collected from multiple electrodes are the linear sum of independent components originating from spatially fixed or overlapping brain regions, whose time delays are negligible. For example, eye artefacts and EEG are independent because their signal production mechanism is entirely different (Onton et al. 2006).

Several unnecessary features take place in the dataset, and the user may not be able to identify the relative importance of these input in terms of classification accuracy. The feature selection methods aim to increase the classification performance by weighting the most contributive parameters to the model performance in the input data. At the same time, the inputs that provide less biomarker information is removed from the input data.

The dominant parts enhancing the classification performance in the input data are commonly selected with the nature-inspired metaheuristic optimisation algorithms (Genetic algorithm, Ant Colony Optimization, Particle Swarm Optimization, etc.) (Cîmpanu et al. 2017; Saeys, Inza, and Larrañaga 2007; Schroder et al. 2003; Zhao, Zhang, and Lei 2014). Generally speaking, the feature selection algorithms work by evaluating a score for each feature and select the best-scored features by considering the results of the trained classifier model.

Classification

The ideal criterion to be used in the classification is to minimise the probability of misclassification. Classification methods (ANN, SVM, Bayesian classifier, Linear Discriminant Analysis (LDA), Hidden Markov Model (HMM), k-Nearest Neighbor, etc.) are used to separate the feature vectors over distinct categories. A classifier should cope with changing patterns, avoid overfitting, and meet the real-time constraints to achieve the classification task (Oskoei and Hu 2007). For example, SVM performs the hyperspace decomposition on linearly indistinguishable data and expands the elements of estimation by mapping them into a high-dimensional space called "feature space".

To find the perfect location for the decision plane is an optimisation-based problem necessitating the kernel function, which establishes linear boundaries via nonlinear transformations (Elhadef 2012; El-Kafrawy, Hegazy, and Tolba 2014). A classifier taxonomy should be made to understand the classification process better. Generative (informative) classifiers compute the likelihood of each class and select most likely (e.g., Bayes quadratic), and discriminative ones learn from the class diversities (e.g., SVM). As static classifiers omit the temporal information (e.g., Multilayer perceptrons, dynamic classifiers capture the temporal dynamics of the feature vectors (e.g., HMM). Stable classifiers are not affected by small variations in the training set (e.g., LDA).

In contrast, unstable classifiers change regarding small variations of the training set (e.g., ANN). Regularised classifiers are robust to outliers, and have increasing generalisation performance. While the dimensionality of the feature vectors increases, the amount of data required to good model the different classes exponentially increases. At least, ten times as many training samples per class (dimensionality) should be utilized (Jain and Chandrasekaran 1982).

The classification aims to find the true label y^* of a feature vector x with a nonlinear transformation mapping f. The mapping model is structured from a training set N. The classification errors are defined over three terms as given in Eq. 1.

$$Error = E\left[\left(y^* - f(x)\right)^2\right] = Noise^2 + Bias(f(x))^2 + Variance(f(x)) \quad (1)$$

where Noise denotes the noise occurred in the system, Bias corresponds to the divergence among the predicted mapping and best mapping. Bias is related to the type of mapping function. Variance means the sensitivity to the training set.

Bias and Variance should be restrained to a lower value. There exists a trade-off situation because of stable classifiers have high bias and low variance and, vice versa for unstable classifiers.

DEEP LEARNING-BASED EEG-BCI

DL algorithms perform better classification, and regression accuracies as compared to the conventional machine learning algorithms for EEG-BCI application since EEG data has high dimensionality, inherently noisy and non-stationary nature, coarse spatial resolution (several centimetres). The task-dependent or event-related sources are affected by noise at a dangerous level. EEG signals consist of a dynamic time series acquired from the multi-channel neuroheadset, whose electrodes surround the scalp surface depicted in Figure 2.

Assuming that an EEG dataset related to a subject is denoted as "D^i". Time segments of a specific class are separated as $D^i = \{(X^1, y^1), \ldots, (X^N, y^N)\}$, where N is the total number of trials for subject i. The input matrix $X^j \in \mathbb{R}^{E.T}$ of trial j, $1 \leq j \leq N_i$ consists of the preprocessed signals of $E - recorded$ electrodes and $T - discretized$ time steps obtained per trial. The class label corresponds to y^j. The classifier is trained to map the label (L) y^j to trial X^j through a nonlinear function $f(X^j; \phi): \mathbb{R}^{E.T} \xrightarrow{with\ parameters\ \phi} L$. $f(X^j; \phi)$ comprises vector-valued feature representation $\Theta(X^j; \phi_\Theta)$ learned from the data, and a classifier ς with parameters ϕ_ς given in Eq. 2 (Schirrmeister et al. 2017)

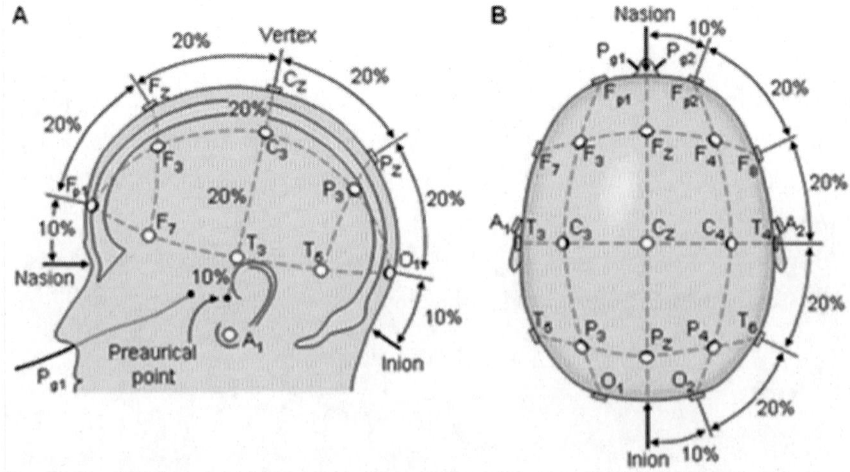

Figure 2. Electrode locations according to the 10/10 International System Positions (John n.d., 20).

$$f\left(X^j, \phi\right) = \varsigma(\Theta\left(X^j; \phi_\Theta\right), \phi_\varsigma) \tag{2}$$

The oscillatory EEG data is decoded via segmentation, detrending, bad trial removal, bad channel removal, and spatio-spectral filtering methodology (e.g., Filter Bank Common Spatial Patterns [FBCSP], Bandpass filtering, STFT, Wavelet Packet Decomposition), which extracts class-separative band power features (Lemm et al. 2005; Paez 2018). A spatial whitening filter is utilised to minimise noise originating from source mixing and volume conduction by re-weighting the electrodes and mapping raw electrode readings to a new space. The sensors in the new space are uncorrelated and have unit power. It is assumed that; frequency band and time window are known, the band-passed signal is jointly Gaussian within the time window, and source activity constellation differs among classes. The spatial filters are designed such that the variance of the filtered signals is maximal for one class while minimal for the other class, and vice versa. The unwanted frequency contents outside the frequency range of interest are eliminated by spectral filtering methodology (Abibullaev and Zollanvari 2019).

The input of the EEG data should be rearranged before feeding the DL network to yield an effective classification performance. The temporal features are extracted via Filter-Bank Common Spatial Patterns (FBCSP) to optimise the Convolutional Neural Network (CNN) performance (Sakhavi, Guan, and Yan 2018). Power Spectral Density (PSD), wavelet decomposition, spectrograms are effective time-frequency input formulations since EEG data is associated with spectral behavioural patterns. After that, EEG data is transformed into the image-based format (Spectrogram images, Fourier feature maps [Tabar and Halici, 2016], designing 2D or 3D grids (Hussein et al. 2018), colour scale topography), and then fed into the CNN with the single or multi-frame approach. The input formulations are depicted in Figure 3.

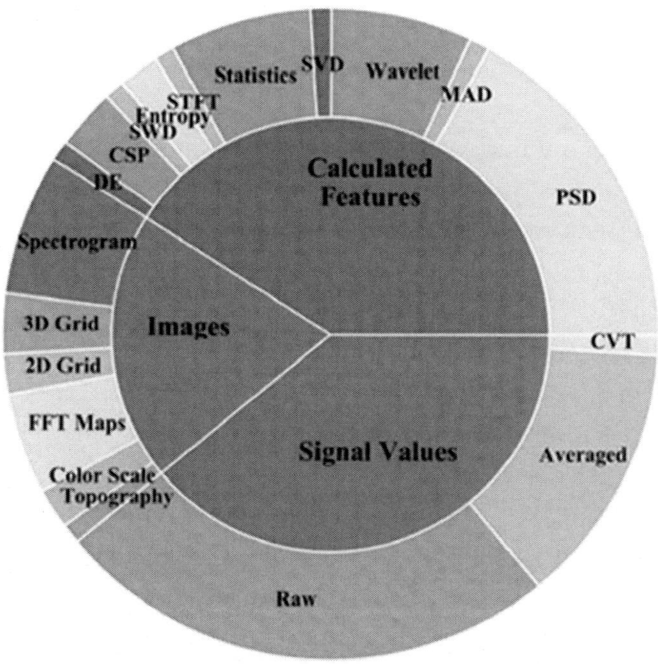

Figure 3. Input formulation methods. CVT: Complex value transformation, CSP: common spatial pattern, DE: dynamic energy, FFT: fast Fourier transform, MAD: mean absolute difference, PSD: power spectral density, STFT: short-time Fourier transformation, SVD: singular value decomposition, SWD: swarm decomposition (Craik, He, and Contreras-Vidal 2019).

The temporal dimension is reduced, and the classes are discriminated through a common convolutional kernel. The classification performance is also optimised by adjusting the hyperparameters like pooling layer number, convolutional layer number, kernel size, dropout and hidden node numbers, respectively. A Deep Recurrent CNN is used to learn EEG representations considering robustness to spatial, spectral, and temporal variations. This model provides a user-independent solution, whose representations do not change in case of various subjects or various occasions for the same subject. Activation functions implemented into the convolutional layers are classified as Rectified Linear Unit (ReLU), Exponential Linear Unit (ELU), Leaky ReLU, Hyperbolic tangent (tanh), scaled ELU, split tanh, parametric ReLU, Softmax, Sigmoid respectively.

Deep Learning Architectures

DL strategies are mainly categorised as Convolutional Neural Network (CNN), Deep Belief Network (DBN), Recurrent Neural Network (RNN), Stacked auto-encoders (SAE), and hybrid architectures, respectively.

Autoencoders

AE is a unique NN structure, which regulates the weights of hidden layer by training to provide the input and output value to be matched. AEs include two data processing phase named as encoding and decoding. Input data $x \in [0 \quad 1]^n$ is fed to the AE during the encoding phase. After that, the hidden layer is constructed regarding encoding function $h = encoder\,(x)\;(h \in [0 \quad 1]^m$. The encoding function is represented as in Eq. 3

$$h = encoder\,(x) = g(W.x + b) \tag{3}$$

where $W \in \mathbb{R}^{m \times n}$ is the weight matrix linking the input layer and the hidden layer. $b \in [0 \quad 1]^m$ is the bias vector, and g denotes the activation function.

Decoding function is given as in Eq. 4

$$y = decoder\ (x) = g(W'.x + b') \tag{4}$$

where $W' \in \mathbb{R}^{n \times m}$ is the weight matrix among hidden layer and output layer, $b \in [0 \quad 1]^n$ is the bias vector.

Model training fits the original input signal to the output signal considering the objective function

$$\min \sum \left| y^{(i)} - x^{(i)} \right|$$

The AEs framework is depicted in Figure 4.

Deep Belief Network

DBN is a non-convolutional, generative and probabilistic model with several layers of latent variables, which are typically binary. Usually, units in each layer are connected to units in each neighbouring layer. Restricted Boltzmann Machine (RBM) is a shallow DBN having a single hidden layer. RBMs are modelled with a Markov random field that builds data distribution by parameterising it through Gibbs method over a pattern among visible "v" and hidden variables "h" given in Figure 5 (Plis et al. 2014).

An RBM comprises an energy-like function for a visible and hidden vector pair $\langle v, h \rangle$ with a weight matrix W related to the connection given in Eq. 5.

$$E(v, h) = -a^T v - b^T h - v^T W h \tag{5}$$

where a and b are the bias weights for visible units and hidden units, respectively. Probability distributions are stated regarding energy function as in Eq. 6.

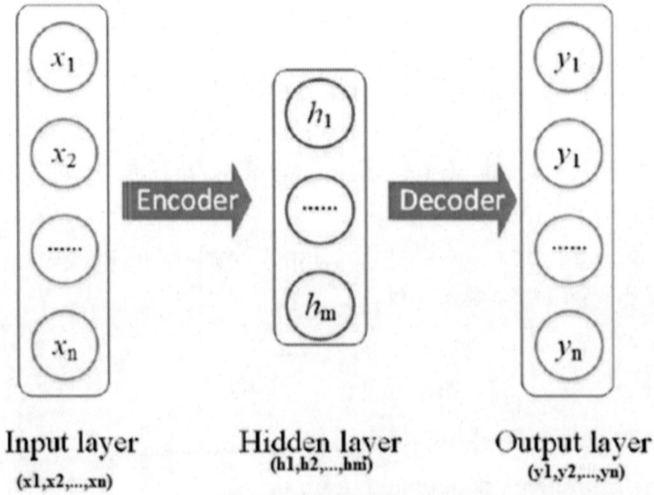

Figure 4. Structure of the autoencoders (Wen and Zhang 2018).

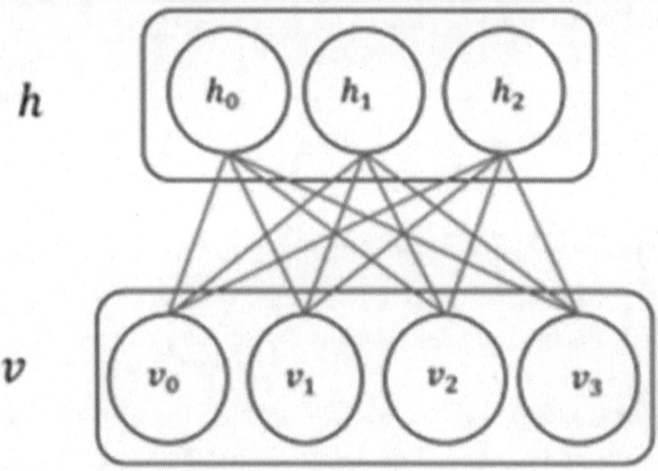

Figure 5. RBM model, including one-layer o hidden units and one layer of visible units (Movahedi, Coyle, and Sejdić 2017).

$$P(v,h) = \frac{1}{Z} e^{-E(v,h)} \tag{6}$$

where Z is a normalizing constant $Z = \sum_{v',h'} e^{-E(v',h')}$

The probability vector of the visible units is defined over hidden units as in Eq. 7.

$$P(v) = \frac{1}{Z}\sum_h e^{-E(v,h)} \tag{7}$$

The derivative of a log-likelihood of training data concerning W is evaluated as in Eq. 8.

$$\sum_{n=1}^{n=N} \frac{\partial logP(v^n)}{\partial W_{ij}} = \langle v_i, h_j \rangle_{data} - \langle v_i, h_j \rangle_{model} \tag{8}$$

where $\langle . \rangle_{data}$ *and* $\langle . \rangle_{model}$ highlight expected values in the data or model distribution. Learning rules for weights is calculated as in Eq. 9.

$$\Delta W_{ij} = \varepsilon\left[\langle v_i, h_j \rangle_{data} - \langle v_i, h_j \rangle_{model}\right] \tag{9}$$

where ε denotes to the learning rate.

Recurrent Neural Network

RNNs are a type of neural networks to process variable-length sequential data. An RNN has a recurrent hidden state. The current activation of this state depends on the previous time step.

The sequence is given as $x = (x_1, x_2, \dots, x_T)$, at each time step, RNN changes its recurrent hidden state "h_t" based on the current input vector "x_t" and the previous hidden state "h_{t-1}" as given in Eq. 10

$$h_t = \begin{cases} 0 \; if \; t = 0 \\ g(h_{t-1}, x_t) \; if \; otherwise \end{cases} \tag{10}$$

where g is the nonlinear function

h_t is updated as $h_t = f(Wx_t + Uh_{t-1} + b)$

where W, U are weights related to input state and previous hidden state, respectively, and b is the bias vector.

$f(.)$ is a pointwise nonlinear activation function (e.g., hyperbolic tangent, logistic sigmoid), whose gradients can grow or decay exponentially during training over time iterations. Affine transformation-based models are applied by using gating units such as e.g., Long Short-term Memory (LSTM), Gated Recurrent Unit (GRU) to design more complex recurrent units (Roy, Kiral-Kornek, and Harrer 2018). The RNN network scheme is represented in Figure 6.

Convolutional Neural Network

CNN's are a type of feed-forward NN, which pass an input more than one hidden layers of neurons representing a linear combination of its input. These combinations are subjected to a nonlinear activation function and passed to the next layers to yield output. The learning ability can be improved by increasing depth-or equivalently hidden layer number. Different from Multi-Layer Perceptrons (MLPs), the weights of the CNN are sparsely-connected, therefore CNN deal with the huge amount and different size of data by processing efficient way. CNN's are resistant and robust to small transmutations in the input data (e.g., scaling, skewing, translation, distortion). Each neuron in CNN includes 2-D kernel for weights called a feature map. The illustration of a basic CNN is given in Figure 7.

Let us assume that; a $N \times N$-pixed grayscale image is the input layer. A linear convolution among the input image and corresponding filter is applied by each neuron in the first convolution layer for producing the input feature map. After that, neurons in the input feature map are subjected to the activation function to yield an output. The pooling process, which is a non-linear downsampling transformation action, performs a decimation in the output feature map of the previous neurons in the first convolutional layer. The first two steps are repeated in the same sequence and the output of the second pooling layer is fed to the fully-connected layers (identical to the MLP). The final output corresponds to the classification information of the input image. CNNs are trained through Backpropagation (BP) algorithm in a supervised way. The max pooling layer scans through the inputs along with a sliding window having a predetermined stride and select the maximum value in every scanned sub-region. The spatial size of the input EEG features is reduced with this operation.

In summary, the EEG-BCI features related to temporal, spatial and spectral domains can be learned through these deep architectures. The comparison of classification accuracies of the DL-based EEG-BCI architectures surveyed in this review is tabulated in Table 1.

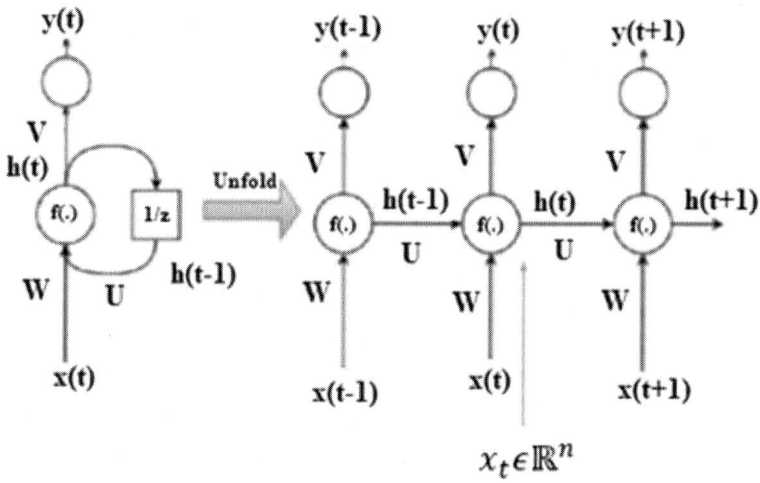

Figure 6. A typical RNN scheme.

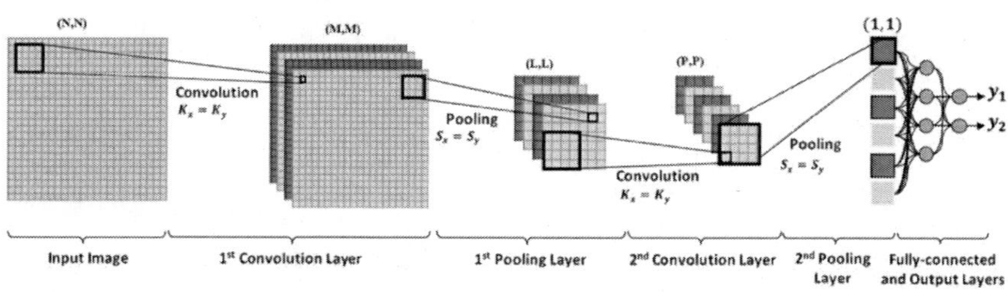

Figure 7. The representation of a basic CNN with two convolutions and one fully-connected layer (Kiranyaz et al. 2019).

Table 1. Comparison of deep learning architecture and input data features across the literature surveyed

Architecture	Input formulation	Modality	Application	Accuracy (%)	
CNN (PodNet: 5 Conv .+ 4 Dropout + 5 Batch Norm. +4 ReLU+ 5 Max.Pool) + Softmax	Downsampling+Normalization	Steady State Visual Evoked Potentials	BCI Speller	86	(Podmore et al. 2019)
Autoencoders + Softmax Layer	CSP	ERS/ERD	Motor Imagery	-	(Kumar et al. 2016)
Parallel Conv.+Average Pool.+ Conv.+ MLP (ReLU)	Filter-Bank CSP	ERS/ERD	Motor Imagery	70	(Sakhavi, Guan, and Yan 2015)
DBN (Restricted Boltzmann Machine + Constrastive Divergence)	FFT	ERS/ERD	Motor Imagery	96	(Shen et al. 2017)
CNN (40 Conv. X2 + Avg.pool + Flatten + 80 Fully Connected +Softmax)	CSP	ERS/ERD	Motor Imagery	86.13	(Dose et al. 2018)
CNN (Conv.+ ReLU + Pool.+ FC + Softmax)	Connectivity Patterns (2D Feature Maps)	ERS/ERD	Motor Imagery	96.69	(Mousapour et al. 2018)
Frequential DBN	FFT + Wavelet Package Decomposition	ERS/ERD	Motor Imagery	71	(Lu et al. 2016)
CNN (Conv + Max.Pool + FC)	CSP, Butterworth LP, Logistic Regression	ERS/ERD	Motor Imagery	91	(Singhal et al. 2019)
EEG2Code CNN Topology	Raw Data	ERP (P300)	BCI Speller	98.7	(Nagel and Spüler 2019)
CNN (Conv.+ Pool.+ Deep MF Softmax)	Scaling (z-score)	ERS/ERD	Motor Imagery	100	(Jingwei, Yin, and Weidong 2015)
Parallel (CNN + Recurrent Neural Network (RNN))	Channel Cross-covariance	ERP	Speech Imagery	90.7	(Saha and Fels 2019)
CNNs-eegNet	Frequency complementary feature map	ERS/ERD	Motor Imagery	69.27	(H. Yang et al. 2015)
DNN (ReLU + Softmax)	Raw Data	EEG+fNIRS	Motor Imagery	95	(Chiarelli et al. 2018)
CNN (Conv. + FC)	Raw Data	ERP (P300)	BCI Speller	87.5	(Cecotti and Graser 2010)
CNN (Conv.+ Pool.+ FC)	Filtering	ERP (P300)	BCI Speller	93.48	(Manor and Geva 2015)

Architecture	Input formulation	Modality	Application	Accuracy (%)	
CNN (Conv2D + DepthwiseConv2D + SeperableConv2D)	FBCSP	ERP/ERN/MRCP/SMR	BCI Speller	-	(Lawhern et al. 2018)
Bayesian Optimization + CNN (Conv.+ Softmax Layer + ReLU)	DFCSP	ERS/ERD	Motor Imagery	85.44	(Olivas-Padilla and Chacon-Murguia 2019)
Stacked Auto-encoder + CNN	FFT	EEG+ECoG	Seziure Prediction and Localization	96	(Hosseini et al. 2017)
AlexNet, VGGnET-16, VGGNet-19, GoogleNet, ResNet-101	KNN regressor	ERP	Object Classification	97.7	(Jiang, Fares, and Zhong 2019)
CNN (Conv.+ Sub-sample + ReLU + Softmax)	Wavelet Packet Decomposition	ERS/ERD	Motor Imagery	78	(Y. Zhang et al. 2019)
CNN+Variational autoencoder	STFT	ERS/ERD	Motor Imagery	-	(Dai et al. 2019)
5-layer CNN (FC)	Time-frequency maps	ERS/ERD	Motor Imagery	86.41	(Tang, Li, and Sun 2017)
Cascade and Parallel Convolutional Recurrent Neural Network	Spatio-temporal maps	Multiple Neuromechanisms	Intention recognition	93	(Q. Zhang and Liu 2018)
deep ConvNet	FBCSP	Multiple Neuromechanisms	Decoding and Visualization	78.5	(Schirrmeister et al. 2017)
Deep Belief Network (Softmax)	Wavelet Packet Transform	ERS/ERD	Motor Imagery	93.57	(Li, Zhang, and Sun 2016)
CNN + SAE	STFT	ERS/ERD	Motor Imagery	85	(Shen et al. 2017)
CNN (Conv.layer + Max pool.+ FC)	Filtering + Normalization	ERS/ERD	Neuro-rehabilitation	78.6	(Karácsony et al. 2019)
Deep Convolutional Generative Adversarial Networks	Complex Morlet Wavelet Transform	VEPs	Improving BCI performance by data augmentation	-	(Q. Zhang and Liu 2018)
Gaussian Bernoulli RBM + PCSA	Raw Data	ECoG	Target prior with deep feature learning	-	(Du et al. 2018)

DISCUSSION

Deep learning has enhanced BCI through end-to-end learning, which means learning from raw data without the necessity of any apriori feature selection. The classification accuracy can be boosted using advanced techniques in deep learning such as dropout, regularisation strategies, batch normalisation, exponential linear units, cropped training strategy, etc. The performance of the BCI is also increased by transforming input data via spatio-spectral modulation, filter bank CSP, FFT maps, meta-heuristic optimisation algorithms, etc.

Through these improvements, the topography of the causal contributions of features in various frequency bands and spatial locations are revealed. The DL methods scale datasets and learn hierarchical structural representative features from data. The disadvantages of DL comprise that may produce false predictions with high confidence [Szegedy et al., 2014] and may necessitate a tremendous amount of training data and hyperparameters. The training time of models last longer, and DL models have lack of interpretability. Despite the disadvantages, these models can be done more interpretable and morefaster through advanced visualisation techniques, and powerful GPUs, respectively. Design choices (e.g., network topology, types of nonlinearities) and training strategies (e.g., trial segmentation, cropping style, shifting window through the trials) determine the performance of the DL methods. The architectures can be adapted to specific needs forced by the multi-channel EEG data analysis.

Thanks to the feature visualisation (e.g., EEG band power features, brain mapping) discriminating the classes from the brain signal activities is also possible to understand the relations among the brain areas and functional connectivity in a neuroscientific manner. DL model behaviour in decoding EEG signals are also tracked with this way. BCI researchers have adopted a variety of DL classifiers, such as RBMs, CNN's, GANs, SAEs. We hope that our literature review will provide BCI researchers with beneficial knowledge in the subject of DL-based EEG-BCIs.

REFERENCES

Abibullaev, Berdakh. & Amin, Zollanvari. (2019). 'Learning Discriminative Spatiospectral Features of ERPs for Accurate Brain-Computer Interfaces'. *IEEE Journal of Biomedical and Health Informatics*.

Alkawadri, Rafeed. (2019). 'Brain Computer Interface (BCI) Applications in Mapping of Epileptic Brain Networks Based on Intracranial-EEG'. *Frontiers in Neuroscience*, *13*, 191.

Artoni, Fiorenzo., Arnaud, Delorme. & Scott, Makeig. (2018). 'Applying Dimension Reduction to EEG Data by Principal Component Analysis Reduces the Quality of Its Subsequent Independent Component Decomposition'. *NeuroImage*, *175*, 176–187.

Birbaumer, Niels., Thilo, Hinterberger., Andrea, Kubler. & Nicola, Neumann. (2003). 'The Thought-Translation Device (TTD): Neurobehavioral Mechanisms and Clinical Outcome'. *IEEE Transactions on Neural Systems and Rehabilitation Engineering*, *11* (2), 120–123.

Braack, Esther M. ter., Benjamin, de Jonge. & Michel, JAM van Putten. (2013). 'Reduction of TMS Induced Artifacts in EEG Using Principal Component Analysis'. *IEEE Transactions on Neural Systems and Rehabilitation Engineering*, *21* (3), 376–382.

Bugli, Céline. & Philippe, Lambert. (2007). 'Comparison between Principal Component Analysis and Independent Component Analysis in Electroencephalograms Modelling'. *Biometrical Journal: Journal of Mathematical Methods in Biosciences*, *49* (2), 312–327.

Caesarendra, Wahyu. (2017). 'A Method to Extract P300 EEG Signal Feature Using Independent Component Analysis (ICA) for Lie Detection'. *Journal of Energy, Mechanical, Material and Manufacturing Engineering*, *2* (1), 9–16.

Castermans, Thierry., Matthieu, Duvinage., Guy, Cheron. & Thierry, Dutoit. (2014). 'Towards Effective Non-Invasive Brain-Computer Interfaces Dedicated to Gait Rehabilitation Systems'. *Brain Sciences*, *4* (1), 1–48.

Cecotti, Hubert. & Axel, Graser. (2010). 'Convolutional Neural Networks for P300 Detection with Application to Brain-Computer Interfaces'. *IEEE Transactions on Pattern Analysis and Machine Intelligence*, *33* (3), 433–445.

Chae, Yongwook., Jaeseung, Jeong. & Sungho, Jo. (2012). 'Toward Brain-Actuated Humanoid Robots: Asynchronous Direct Control Using an EEG-Based BCI'. *IEEE Transactions on Robotics*, *28* (5), 1131–1144.

Chiarelli, Antonio Maria., Pierpaolo, Croce., Arcangelo, Merla. & Filippo, Zappasodi. (2018). 'Deep Learning for Hybrid EEG-FNIRS Brain–Computer Interface: Application to Motor Imagery Classification'. *Journal of Neural Engineering*, *15* (3), 036028.

Cîmpanu, Corina., Lavinia, Ferariu., Tiberius, Dumitriu. & Florina, Ungureanu. (2017). 'Multi-Objective Optimization of Feature Selection Procedure for EEG Signals Classification'. In *2017 E-Health and Bioengineering Conference (EHB)*, 434–437. IEEE.

Cole, Scott R. & Bradley, Voytek. (2018). 'Cycle-by-Cycle Analysis of Neural Oscillations'. *BioRxiv*, *302000*.

Cong, Fengyu., Vinoo, Alluri., Asoke, K. Nandi., Petri, Toiviainen., Rui, Fa., Basel, Abu-Jamous., Liyun, Gong., Bart, G. W. Craenen., Hanna, Poikonen. & Minna, Huotilainen. (2013). 'Linking Brain Responses to Naturalistic Music through Analysis of Ongoing EEG and Stimulus Features'. *IEEE Transactions on Multimedia*, *15* (5), 1060–1069.

Costa, Álvaro., Enrique, Hortal., Eduardo, Iáñez. & José, M. Azorín. (2014). 'A Supplementary System for a Brain-Machine Interface Based on Jaw Artifacts for the Bidimensional Control of a Robotic Arm'. *PloS One*, *9* (11), e112352.

Coyle, Damien., Girijesh, Prasad. & Martin McGinnity, T. (2005). 'A Time-Frequency Approach to Feature Extraction for a Brain-Computer Interface with a Comparative Analysis of Performance Measures'. *EURASIP Journal on Applied Signal Processing*, 3141–3151.

Craik, Alexander., Yongtian, He. & Jose, Luis Pepe Contreras-Vidal. (2019). 'Deep Learning for Electroencephalogram (EEG) Classification Tasks: A Review'. *Journal of Neural Engineering*.

Dai, Mengxi., Dezhi, Zheng., Rui, Na., Shuai, Wang. & Shuailei, Zhang. (2019). 'EEG Classification of Motor Imagery Using a Novel Deep Learning Framework'. *Sensors*, *19* (3), 551.

Daly, Ian., Nicoletta, Nicolaou., Slawomir, Jaroslaw Nasuto. & Kevin, Warwick. (2013). 'Automated Artifact Removal from the Electroencephalogram: A Comparative Study'. *Clinical EEG and Neuroscience*, *44* (4), 291–306.

Das, Rig., Emanuele, Maiorana. & Patrizio, Campisi. (2017). 'Visually Evoked Potential for EEG Biometrics Using Convolutional Neural Network'. In *2017 25th European Signal Processing Conference (EUSIPCO)*, 951–955. IEEE.

Dasgupta, Saumitro., Michael, Fanton., Jonathan, Pham., Michael, Willard., Nezamfar, H., Shafai, B. & Erdogmus, D. (2010). 'Brain Controlled Robotic Platform Using Steady State Visual Evoked Potentials Acquired by EEG'. In *2010 Conference Record of the Forty Fourth Asilomar Conference on Signals, Systems and Computers*, 1371–1374. IEEE.

Dose, Hauke., Jakob, S. Møller., Sadasivan, Puthusserypady. & Helle, K. Iversen. (2018). 'A Deep Learning MI-EEG Classification Model for BCIs'. In *2018 26th European Signal Processing Conference (EUSIPCO)*, 1676–1679. IEEE.

Du, Anming., Shuqin, Yang., Weijia, Liu. & Haiping, Huang. (2018). 'Decoding ECoG Signal with Deep Learning Model Based on LSTM'. In *TENCON 2018-2018 IEEE Region 10 Conference*, 0430–0435. IEEE.

Elhadef, Mourad. (2012). 'Using Linear Support Vector Machines to Solve the Asymmetric Comparison-Based Fault Diagnosis Problem'. In *2012 Seventh International Conference on Availability, Reliability and Security*, 18–27. IEEE.

El-Kafrawy, Noran M., Doaa, Hegazy. & Mohamed, Fahmy Tolba. (2014). 'Features Extraction and Classification of Eeg Signals Using Empirical Mode Decomposition and Support Vector Machine'. In *International Conference on Advanced Machine Learning Technologies and Applications*, 189–198. Springer.

Eom, Jin-Sup., Hye-Ryeon, Yang., Mi-Sook, Park. & Jin-Hun, Sohn. (2013). 'P300 Speller Using a New Stimulus Presentation Paradigm'. In *2013 International Winter Workshop on Brain-Computer Interface (BCI)*, 98–99. IEEE.

Fortino, Giancarlo., Giuseppe, Di Fatta., Mukaddim, Pathan. & Athanasios, V. Vasilakos. (2014). 'Cloud-Assisted Body Area Networks: State-of-the-Art and Future Challenges'. *Wireless Networks*, *20* (7), 1925–1938.

Gao, Shangkai., Yijun, Wang., Xiaorong, Gao. & Bo, Hong. (2014). 'Visual and Auditory Brain–Computer Interfaces'. *IEEE Transactions on Biomedical Engineering*, *61* (5), 1436–1447.

Goodfellow, Ian., Yoshua, Bengio. & Aaron, Courville. (2016). *Deep Learning*. MIT press.

Gope, C., Kehtarnavaz, N. & Nair, D. (2005). 'Neural Network Classification of EEG Signals Using Time-Frequency Representation'. In *Proceedings. 2005 IEEE International Joint Conference on Neural Networks*, *4*, 2502–2507. IEEE.

Guger, Christoph., Herbert, Ramoser. & Gert, Pfurtscheller. (2000). 'Real-Time EEG Analysis with Subject-Specific Spatial Patterns for a Brain-Computer Interface (BCI)'. *IEEE Transactions on Rehabilitation Engineering*, *8* (4), 447–456.

Hortal, Enrique., Daniel, Planelles., Costa, A., Eduardo, Iáñez., Andrés, Úbeda., José, Maria Azorín. & Eduardo, Fernández. (2015). 'SVM-Based Brain–Machine Interface for Controlling a Robot Arm through Four Mental Tasks'. *Neurocomputing*, *151*, 116–121.

Hosseini, Mohammad-Parsa., Dario, Pompili., Kost, Elisevich. & Hamid, Soltanian-Zadeh. (2017). 'Optimized Deep Learning for EEG Big Data and Seizure Prediction BCI via Internet of Things'. *IEEE Transactions on Big Data*, *3* (4), 392–404.

Hussein, Ramy., Hamid, Palangi., Jane Wang, Z. & Rabab, Ward. (2018). 'Robust Detection of Epileptic Seizures Using Deep Neural Networks'. In *2018 IEEE International Conference on Acoustics*, *Speech and Signal Processing (ICASSP)*, 2546–2550. IEEE.

Iba, Hitoshi. (2018). 'Meta-Heuristics, Machine Learning, and Deep Learning Methods'. In *Evolutionary Approach to Machine Learning and Deep Neural Networks*, 27–75. Springer.

Inoue, Shinsuke., Yoko, Akiyama., Yoshinobu, Izumi. & Shigehiro, Nishijima. (2008). 'The Development of BCI Using Alpha Waves for Controlling the Robot Arm'. *IEICE Transactions on Communications*, *91* (7), 2125–2132.

Iturrate, Inaki., Javier, Antelis. & Javier, Minguez. (2009). 'Synchronous EEG Brain-Actuated Wheelchair with Automated Navigation'. In *2009 IEEE International Conference on Robotics and Automation*, 2318–2325. IEEE.

Jain, Anil K. & Balakrishnan, Chandrasekaran. (1982). '39 Dimensionality and Sample Size Considerations in Pattern Recognition Practice'. *Handbook of Statistics*, *2*, 835–855.

Jiang, Jianmin., Ahmed, Fares. & Sheng-Hua, Zhong. (2019). 'A Context-Supported Deep Learning Framework for Multimodal Brain Imaging Classification'. *IEEE Transactions on Human-Machine Systems.*

Jingwei, Liu., Cheng, Yin. & Zhang, Weidong. (2015). 'Deep Learning EEG Response Representation for Brain Computer Interface'. In *2015 34th Chinese Control Conference (CCC)*, 3518–3523. IEEE.

John. n.d. '10/20 System Positioning | DIY TDCS'. Accessed 22 June 2019. https://www.diytdcs.com/tag/1020-system-positioning/.

Kaplan, A. Ya. & Sergei, L. Shishkin. (2000). 'Application of the Change-Point Analysis to the Investigation of the Brain's Electrical Activity'. In *Non-Parametric Statistical Diagnosis*, 333–388. Springer.

Karácsony, Tamás., John, Paulin Hansen., Helle, Klingenberg Iversen. & Sadasivan, Puthusserypady. (2019). 'Brain Computer Interface for Neuro-Rehabilitation With Deep Learning Classification and Virtual Reality Feedback'. In *10th Augmented Human International Conference 2019*, Article–No. Association for Computing Machinery.

Khushaba, Rami N., Sarath, Kodagoda., Sara, Lal. & Gamini, Dissanayake. (2010). 'Driver Drowsiness Classification Using Fuzzy Wavelet-Packet-Based Feature-Extraction Algorithm'. *IEEE Transactions on Biomedical Engineering*, *58* (1), 121–131.

Kim, Yoon Jae., Sung, Woo Park., Hong, Gi Yeom., Moon, Suk Bang., June, Sic Kim., Chun, Kee Chung. & Sungwan, Kim. (2015). 'A Study on a Robot Arm Driven by Three-Dimensional Trajectories Predicted from Non-Invasive Neural Signals'. *Biomedical Engineering Online*, *14* (1), 81.

Kiranyaz, Serkan., Onur, Avci., Osama, Abdeljaber., Turker, Ince., Moncef, Gabbouj. & Daniel, J. Inman. (2019). '1D Convolutional Neural Networks and Applications: A Survey'. *ArXiv Preprint ArXiv*, 1905.03554.

Kline, Jacob. (2012). *Handbook of Biomedical Engineering*. Elsevier.

Ku, Y. G. & Masashi, Kawasumi. (2007). 'Nonstationary EEG Analysis Using Random-Walk Model'. In *World Congress on Medical Physics and Biomedical Engineering 2006*, 1067–1070. Springer.

Kumar, Shiu., Alok, Sharma., Kabir, Mamun. & Tatsuhiko, Tsunoda. (2016). 'A Deep Learning Approach for Motor Imagery EEG Signal Classification'. In *2016 3rd Asia-Pacific World Congress on Computer Science and Engineering (APWC on CSE)*, 34–39. IEEE.

Lawhern, Vernon J., Amelia, J. Solon., Nicholas, R. Waytowich., Stephen, M. Gordon., Chou, P. Hung. & Brent, J. Lance. (2018). 'EEGNet: A Compact Convolutional Neural Network for EEG-Based Brain–Computer Interfaces'. *Journal of Neural Engineering*, *15* (5), 056013.

LeCun, Yann., Yoshua, Bengio. & Geoffrey, Hinton. (2015). 'Deep Learning'. *Nature*, *521* (7553), 436.

Lemm, Steven., Benjamin, Blankertz., Gabriel, Curio. & Muller, K. R. (2005). 'Spatio-Spectral Filters for Improving the Classification of Single Trial EEG'. *IEEE Transactions on Biomedical Engineering*, *52* (9), 1541–1548.

Li, Ming-Ai., Meng, Zhang. & Yan-Jun, Sun. (2016). 'A Novel Motor Imagery EEG Recognition Method Based on Deep Learning'. In *2016 International Forum on Management, Education and Information Technology Application*. Atlantis Press.

Long, Jinyi., Yuanqing, Li., Hongtao, Wang., Tianyou, Yu., Jiahui, Pan. & Feng, Li. (2012). 'A Hybrid Brain Computer Interface to Control the Direction and Speed of a Simulated or Real Wheelchair'. *IEEE Transactions on Neural Systems and Rehabilitation Engineering, 20* (5), 720–729.

Lu, Na., Tengfei, Li., Xiaodong, Ren. & Hongyu, Miao. (2016). 'A Deep Learning Scheme for Motor Imagery Classification Based on Restricted Boltzmann Machines'. *IEEE Transactions on Neural Systems and Rehabilitation Engineering, 25* (6), 566–576.

Machado, Sergio., Fernanda, Araújo., Flávia, Paes., Bruna, Velasques., Mario, Cunha., Henning, Budde., Luis, F. Basile., Renato, Anghinah., Oscar, Arias-Carrión. & Mauricio, Cagy. (2010). 'EEG-Based Brain-Computer Interfaces: An Overview of Basic Concepts and Clinical Applications in Neurorehabilitation'. *Reviews in the Neurosciences, 21*, (6), 451–468.

Manor, Ran. & Amir, B. Geva. (2015). 'Convolutional Neural Network for Multi-Category Rapid Serial Visual Presentation BCI'. *Frontiers in Computational Neuroscience, 9*, 146.

Meyer, Timm., Jan, Peters., Doris, Brötz., Thorsten, O. Zander., Bernhard, Schölkopf., Surjo, R. Soekadar. & Moritz, Grosse-Wentrup. (2012). 'A Brain-Robot Interface for Studying Motor Learning after Stroke'. In *2012 IEEE/RSJ International Conference on Intelligent Robots and Systems*, 4078–4083. IEEE.

Michel, Christoph M. & Bin, He. (2012). 'EEG Mapping and Source Imaging'. In *Niedermeyer's Electroencephalography: Basic Principles, Clinical Applications, and Related Fields: Sixth Edition*, 1179–1202. Wolters Kluwer Health Adis (ESP).

Mirghasemi, H., Fazel-Rezai, R. & Shamsollahi, M. B. (2006). 'Analysis of P300 Classifiers in Brain Computer Interface Speller'. In *2006 International Conference of the IEEE Engineering in Medicine and Biology Society*, 6205–6208. IEEE.

Moore, Melody M. (2003). 'Real-World Applications for Brain-Computer Interface Technology'. *IEEE Transactions on Neural Systems and Rehabilitation Engineering, 11* (2), 162–165.

Moreno-Valenzuela, Javier. & Adriana, Salinas-Avila. (2011). 'Biomimetic Control of Mechanical Systems Equipped with Musculotendon Actuators'. *Journal of Bionic Engineering*, *8* (1), 56–68.

Mousapour, Leila., Fateme, Agah., Soorena, Salari. & Marzieh, Zare. (2018). 'A Novel Approach to Classify Motor-Imagery EEG with Convolutional Neural Network Using Network Measures'. In *2018 4th Iranian Conference on Signal Processing and Intelligent Systems (ICSPIS)*, 43–47. IEEE.

Movahedi, Faezeh., James, L. Coyle. & Ervin, Sejdić. (2017). 'Deep Belief Networks for Electroencephalography: A Review of Recent Contributions and Future Outlooks'. *IEEE Journal of Biomedical and Health Informatics*, 22 (3), 642–652.

Muehl, Christian., Brendan, Allison., Anton, Nijholt. & Guillaume, Chanel. (2014). *Affective Brain-Computer Interfaces: Special Issue Editorial*. Taylor & Francis.

Muhammed, Mustafa Alghali Elsaid,, Ahmed, Abdalazeem Ahmed. & Tarig, Ahmed Khalid. (2017). 'Benchmark Analysis of Popular ImageNet Classification Deep CNN Architectures'. In *2017 International Conference On Smart Technologies For Smart Nation (SmartTechCon)*, 902–907. IEEE.

Nagel, Sebastian. & Martin, Spüler. (2019). 'World's Fastest Brain-Computer Interface: Combining EEG2Code with Deep Learning'. *BioRxiv*, 546986.

Olivas-Padilla, Brenda E. & Mario, I. Chacon-Murguia. (2019). 'Classification of Multiple Motor Imagery Using Deep Convolutional Neural Networks and Spatial Filters'. *Applied Soft Computing*, *75*, 461–472.

Onton, Julie., Marissa, Westerfield., Jeanne, Townsend. & Scott, Makeig. (2006). 'Imaging Human EEG Dynamics Using Independent Component Analysis'. *Neuroscience & Biobehavioral Reviews*, *30* (6), 808–822.

Oskoei, Mohammadreza Asghari. & Huosheng, Hu. (2007). 'Myoelectric Control Systems—A Survey'. *Biomedical Signal Processing and Control*, *2* (4), 275–294.

Paez, Antonio. (2018). 'Using Spatial Filters and Exploratory Data Analysis to Enhance Regression Models of Spatial Data'. *Geographical Analysis*.

Pan, Jiahui., Qiuyou, Xie., Haiyun, Huang., Yanbin, He., Yuping, Sun., Ronghao, Yu. & Yuanqing, Li. (2018). 'Emotion-Related Consciousness Detection in Patients With Disorders of Consciousness Through an EEG-Based BCI System'. *Frontiers in Human Neuroscience*, *12*, 198.

Philips, Johan., José, del R. Millán., Gerolf, Vanacker., Eileen, Lew., Ferran, Galán., Pierre, W. Ferrez., Hendrik, Van Brussel. & Marnix, Nuttin. (2007). 'Adaptive Shared Control of a Brain-Actuated Simulated Wheelchair'. In *2007 IEEE 10th International Conference on Rehabilitation Robotics*, 408–414. IEEE.

Plis, Sergey M., Devon, R. Hjelm., Ruslan, Salakhutdinov., Elena, A. Allen., Henry, J. Bockholt., Jeffrey, D. Long., Hans, J. Johnson., Jane, S. Paulsen., Jessica, A. Turner. & Vince, D. Calhoun. (2014). 'Deep Learning for Neuroimaging: A Validation Study'. *Frontiers in Neuroscience*, *8*, 229.

Podmore, Joshua J., Toby, P Breckon., Nik, KN Aznan., Jason, D Connolly. & others. (2019). 'On the Relative Contribution of Deep Convolutional Neural Networks for SSVEP-Based Bio-Signal Decoding in BCI Speller Applications.' *IEEE Transactions on Neural Systems and Rehabilitation Engineering: A Publication of the IEEE Engineering in Medicine and Biology Society*.

Portelli, Anthony. & Slawomir, Nasuto. (2017). 'Design and Development of Non-Contact Bio-Potential Electrodes for Pervasive Health Monitoring Applications'. *Biosensors*, *7* (1), 2.

Ramesh, Siliveru., Gopi Krishna, M. & Madhu, Nakirekanti. (2014). 'Brain Computer Interface System for Mind Controlled Robot Using Bluetooth'. *International Journal of Computer Applications*, *104* (15).

Roy, Subhrajit., Isabell, Kiral-Kornek. & Stefan, Harrer. (2018). 'ChronoNet: A Deep Recurrent Neural Network for Abnormal EEG Identification'. *ArXiv Preprint ArXiv*, 1802.00308.

Saeys, Yvan., Iñaki, Inza. & Pedro, Larrañaga. (2007). 'A Review of Feature Selection Techniques in Bioinformatics'. *Bioinformatics*, *23* (19), 2507–2517.

Saha, Pramit. & Sidney, Fels. (2019). 'Hierarchical Deep Feature Learning for Decoding Imagined Speech from Eeg'. *ArXiv Preprint ArXiv*, 1904.04352.

Sakhavi, Siavash., Cuntai, Guan. & Shuicheng, Yan. (2015). 'Parallel Convolutional-Linear Neural Network for Motor Imagery Classification'. In *2015 23rd European Signal Processing Conference (EUSIPCO)*, 2736–2740. IEEE.

———. (2018). 'Learning Temporal Information for Brain-Computer Interface Using Convolutional Neural Networks'. *IEEE Transactions on Neural Networks and Learning Systems*, no. 99, 1–11.

Sakkalis, Vangelis., Tracey, Cassar., Michalis, Zervakis., Kenneth, P. Camilleri., Simon, G. Fabri., Cristin, Bigan., Eleni, Karakonstantaki. & Sifis, Micheloyannis. (2008). 'Parametric and Nonparametric EEG Analysis for the Evaluation of EEG Activity in Young Children with Controlled Epilepsy'. *Computational Intelligence and Neuroscience*, *1*.

Schalk, Gerwin. & Eric, C. Leuthardt. (2011). 'Brain-Computer Interfaces Using Electrocorticographic Signals'. *IEEE Reviews in Biomedical Engineering*, *4*, 140–154.

Schalk, Gerwin., Dennis, J. McFarland., Thilo, Hinterberger., Niels, Birbaumer. & Jonathan, R. Wolpaw. (2004). 'BCI2000: A General-Purpose Brain-Computer Interface (BCI) System'. *IEEE Transactions on Biomedical Engineering*, *51* (6), 1034–1043.

Schirrmeister, R., Lukas, Gemein., Katharina, Eggensperger., Frank, Hutter. & Tonio, Ball. (2017). 'Deep Learning with Convolutional Neural Networks for Decoding and Visualization of EEG Pathology'. In *2017 IEEE Signal Processing in Medicine and Biology Symposium (SPMB)*, 1–7. IEEE.

Schroder, Michael., Martin, Bogdan., Thhilo, Hinterberger. & Niels, Birbaumer. (2003). 'Automated EEG Feature Selection for Brain Computer Interfaces'. In *First International IEEE EMBS Conference on Neural Engineering, 2003. Conference Proceedings.*, 626–629. IEEE.

Schröer, Sebastian., Ingo, Killmann., Barbara, Frank., Martin, Völker., Lukas, Fiederer., Tonio, Ball. & Wolfram, Burgard. (2015). 'An Autonomous Robotic Assistant for Drinking'. In *2015 IEEE International Conference on Robotics and Automation (ICRA)*, 6482–6487. IEEE.

Sharma, Atharva., Xiuwen, Liu. & Xiaojun, Yang. (2018). 'Land Cover Classification from Multi-Temporal, Multi-Spectral Remotely Sensed Imagery Using Patch-Based Recurrent Neural Networks'. *Neural Networks*, *105*, 346–355.

Shen, Peng., Xugang, Lu., Sheng, Li. & Hisashi, Kawai. (2017). 'Conditional Generative Adversarial Nets Classifier for Spoken Language Identification.' In *INTERSPEECH*, 2814–2818.

Shiman, F., Safavi, S. H., Vaneghi, F. M., Oladazimi, M., Safari, M. J. & Ibrahim, F. (2012). 'EEG Feature Extraction Using Parametric and Non-Parametric Models'. In *Proceedings of 2012 IEEE-EMBS International Conference on Biomedical and Health Informatics*, 66–70. IEEE.

Singhal, Kriti., Evi, Agarwal., Arpit, Yadav. & Anuraj, Singh. (2019). 'Classification of Hand Movement Stages for Brain–Computer Interface Using Convolutional Neural Network'. In *Computational Intelligence: Theories, Applications and Future Directions-Volume II*, 189–199. Springer.

Sun, Lisha., Ying, Liu. & Patch, J. Beadle. (2005). 'Independent Component Analysis of EEG Signals'. In *Proceedings of 2005 IEEE International Workshop on VLSI Design and Video Technology*, 219–222. IEEE.

Tan, Desney. & Anton, Nijholt. (2010). 'Brain-Computer Interfaces and Human-Computer Interaction'. In *Brain-Computer Interfaces*, 3–19. Springer.

Tang, Zhichuan., Chao, Li. & Shouqian, Sun. (2017). 'Single-Trial EEG Classification of Motor Imagery Using Deep Convolutional Neural Networks'. *Optik-International Journal for Light and Electron Optics*, 130, 11–18.

Thomas, Eoin., Joan, Fruitet. & Maureen, Clerc. (2012). 'Investigating Brief Motor Imagery for an ERD/ERS Based BCI'. In *2012 Annual International Conference of the IEEE Engineering in Medicine and Biology Society*, 2929–2932. IEEE.

Toderean, Roxana. & Iuliana, Chiuchisan. (2017). 'Application of Support Vector Machine for the Classification of Sensorimotor Rhythms in Brain Computer Interface'. In *2017 E-Health and Bioengineering Conference (EHB)*, 663–666. IEEE.

Treder, Matthias. & Guido, Nolte. (2018). 'Source Reconstruction of Broadband EEG/MEG Data Using the Frequency-Adaptive Broadband (FAB) Beamformer'. *BioRxiv*, 502690.

Trejo, Leonard J., Roman, Rosipal. & Bryan, Matthews. (2006). 'Brain-Computer Interfaces for 1-D and 2-D Cursor Control: Designs Using Volitional Control of the EEG Spectrum or Steady-State Visual Evoked Potentials'. *IEEE Transactions on Neural Systems and Rehabilitation Engineering*, *14* (2), 225–229.

Tzallas, Alexandros T., Vaggelis, P. Oikonomou. & Dimitrios, I. Fotiadis. (2006). 'Epileptic Spike Detection Using a Kalman Filter Based Approach'. In *2006 International Conference of the IEEE Engineering in Medicine and Biology Society*, 501–504. IEEE.

Ullah, Sana., Athanasios, Vasilakos., Han-Chieh, Chao. & Junichi, Suzuki. (2014). 'Cloud-Assisted Wireless Body Area Networks'. *Information Sciences-Special Issue on Cloud-Assisted Wireless Body Area Networks*, 81–83.

Valbuena, Diana., Marco, Cyriacks., Ola, Friman., Ivan, Volosyak. & Axel, Graser. (2007). 'Brain-Computer Interface for High-Level Control of Rehabilitation Robotic Systems'. In *2007 IEEE 10th International Conference on Rehabilitation Robotics*, 619–625. IEEE.

Velasco-Álvarez, Francisco., Álvaro, Fernández-Rodríguez., Antonio, Díaz-Estrella., María, J. Blanca-Mena. & Ricardo, Ron-Angevin.

(2018). 'Control Strategies of a Brain-Controlled Wheelchair Using Two Mental Tasks'. In *Smart Wheelchairs and Brain-Computer Interfaces*, 345–368. Elsevier.

Wen, Tingxi. & Zhongnan, Zhang. (2018). 'Deep Convolution Neural Network and Autoencoders-Based Unsupervised Feature Learning of EEG Signals'. *IEEE Access*, *6*, 25399–25410.

Xingang, Zhao., An, Lipeng. & Ye, Dan. (2014). 'Design of Upper Extremity Motor Function Mirror Rehabilitation System for Post-Stroke Patients Based on FES'. In *The 26th Chinese Control and Decision Conference (2014 CCDC)*, 4606–4611. IEEE.

Yan, Shiyu., Haibin, Zhao., Chong, Liu. & Hong, Wang. (2012). 'Brain-Computer Interface Design Based on Wavelet Packet Transform and SVM'. In *2012 International Conference on Systems and Informatics (ICSAI2012)*, 1054–1056. IEEE.

Yan, Tang., Tang, Jingtian. & Gong, Andong. (2008). 'Multi-Class EEG Classification for Brain Computer Interface Based on CSP'. In *2008 International Conference on BioMedical Engineering and Informatics*, *2*, 469–472. IEEE.

Yang, Chunfeng., Régine, Le Bouquin Jeannès., Jean-Jacques, Bellanger. & Huazhong, Shu. (2012). 'A New Strategy for Model Order Identification and Its Application to Transfer Entropy for EEG Signals Analysis'. *IEEE Transactions on Biomedical Engineering*, *60* (5), 1318–1327.

Yang, Huijuan., Siavash, Sakhavi., Kai, Keng Ang. & Cuntai, Guan. (2015). 'On the Use of Convolutional Neural Networks and Augmented CSP Features for Multi-Class Motor Imagery of EEG Signals Classification'. In *2015 37th Annual International Conference of the IEEE Engineering in Medicine and Biology Society (EMBC)*, 2620–2623. IEEE.

Yang, Yilong., Qingfeng, Wu., Ming, Qiu., Yingdong, Wang. & Xiaowei, Chen. (2018). 'Emotion Recognition from Multi-Channel EEG through Parallel Convolutional Recurrent Neural Network'. In *2018 International Joint Conference on Neural Networks (IJCNN)*, 1–7. IEEE.

Zecca, M., Silvestro, Micera., Carrozza, M. C. & Dario, P. (2017). 'Control of Multifunctional Prosthetic Hands by Processing the Electromyographic Signal'. *Critical Reviews^{TM} in Biomedical Engineering*, *45* (1–6).

Zhang, Qiqi. & Ying, Liu. (2018). 'Improving Brain Computer Interface Performance by Data Augmentation with Conditional Deep Convolutional Generative Adversarial Networks'. *ArXiv Preprint ArXiv*, 1806.07108.

Zhang, Wenchang., Fuchun, Sun., Jianhua, Chen., Chuanqi, Tan., Hang, Wu. & Weihua, Su. (2017). 'An Asynchronous Mi-Based BCI for Brain-Actuated Robot Grasping Control'. In *2017 International Conference on Computer Systems, Electronics and Control (ICCSEC)*, 893–898. IEEE.

Zhang, Yu., Xiong, Zhang., Han, Sun., Zhaowen, Fan. & Xuefei, Zhong. (2019). 'Portable Brain-Computer Interface Based on Novel Convolutional Neural Network'. *Computers in Biology and Medicine*.

Zhang, Zhaoyang., Honggang, Wang., Athanasios, V. Vasilakos. & Hua, Fang. (2012). 'ECG-Cryptography and Authentication in Body Area Networks'. *IEEE Transactions on Information Technology in Biomedicine*, *16* (6), 1070–1078.

Zhao, Xiaoming., Shiqing, Zhang. & Bicheng, Lei. (2014). 'Robust Emotion Recognition in Noisy Speech via Sparse Representation'. *Neural Computing and Applications*, *24* (7–8), 1539–53. https://doi.org/10.1007/s00521-013-1377-z.

In: Human-Computer Interaction
Editor: Turgut Ozseven
ISBN: 978-1-53616-495-4

Chapter 2

A Hybrid Facial Emotion Recognition Framework Using Deep Learning Methodologies

Abdullah Asım Yılmaz, Mehmet Serdar Güzel, İman Askerbeyli and Erkan Bostancı [*]
Department of Computer Engineering, Ankara University, Ankara, Turkey

Abstract

Facial emotion recognition is an important working field in computer vision and artificial intelligence due to its potential in both academic and commercial worlds. That is because, in addition to verbal expressions in interactive communication, facial emotions are also significant factors that help humans to understand the intentions of others. Besides, this working area is also deeply associated with Human computer-interaction (HCI) that essentially examines and facilities how people can interact with machines using their emotions and facial gestures. This study proposes a novel deep neural network architecture using transfer-learning methodology for real-time facial emotion recognition tasks. Essentially, these tasks are performed with high success using the proposed deep

[*] Corresponding Author's Email: ebostanci@ankara.edu.tr.

neural network architecture. One of the main contribution of this study is to recognize facial emotions obtained from real-time video frames with an optimized manner. The classification of emotions is reduced into seven categories, namely, angry, disgust, fear, happiness, sadness, surprise and neutrality. The proposed architecture involves four main stages. These are respectively the creation of the data set, the design of the deep neural network architecture, the training of the proposed deep neural network architecture and the evaluation of the trained deep neural network. Additionally, in order to reveal the superiority of the proposed performance over conventional architectures, performance evaluation between those architectures based on comprehensive datasets are also presented and discussed in this study. Results reveal that the proposed architecture outperforms previously developed architectures in terms of estimating facial emotions and gestures.

Keywords: facial emotion recognition, deep neural networks, computer vision, human computer interaction, deep learning

INTRODUCTION

Facial emotions and gestures are considered as critical factors in terms of helping humans to recognize the intentions of others. Humans often infer sense about their emotional states such as surprise, happiness and sadness with make benefit of the facial expressions and voice tones of other people. According to different research, it was observed that non-verbal components have much more effect than oral components in interpersonal face-to-face communication (Mehrabian 2008; Essa and Pentland 1997; Kaulard et al. 2012; Srivastava 2012). Therefore, the recognition process of facial emotion by the computers in the field of informatics plays a key role in human computer interaction and has been an effective research topic in recent years, which has involved the attention of many researches. Application areas such as human-computer interaction (Dornaika and Raducanu 2007; Bartneck and Lyons 2007), driving-safety (Gao, Yüce, and Thiran 2014), health-care (Muhammad et al. 2017), virtual reality (Hickson et al. 2019), augmented reality (C.-H. Chen, Lee, and Lin 2015), image reconstruction (Mutlu et al. 2014) and so on are examples of application areas where facial emotion recognition operations are mainly employed.

The problem of recognizing facial emotions as a whole can be defined as follows: First, it is the process of data acquisition for a computer system that the facial features from the face area obtained from detected human faces in images or videos, and then the movement of the facial features or gaining the ability capability to interpret this information obtained as a result of analyzing changes in the appearance of facial features as some facial emotions such as smile and anger. Methodologies of recognizing facial emotions are observed in two main groups at the literature depending on whether the attributes are generated by handcrafted (conventional method) (Güzel 2016) or output of a deep neural network model (deep learning-based) (Gao, Yüce, and Thiran 2014).

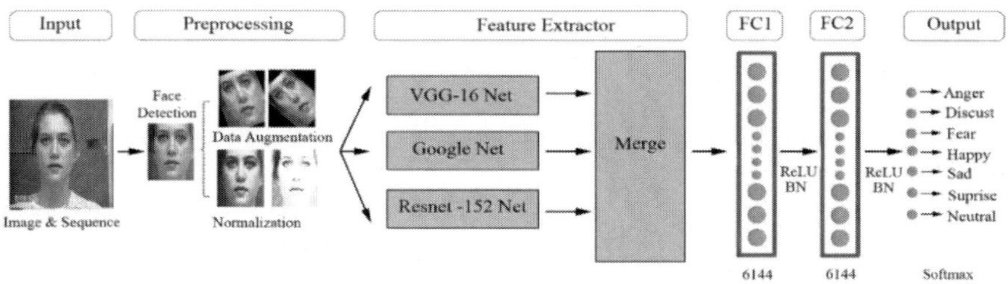

Figure 1. The general pipeline of proposed facial emotion recognition systems.

Within the proposed work, a novel deep neural network architecture is designed and implemented using transfer learning approach for real-time facial emotion recognition tasks, and corresponding operations are conducted by employing this deep neural network architecture. Figure 1 illustrates the proposed pipeline of the system. According to which, first input frames are obtained in real time which is followed by face detection and preprocessing operations. Once this stage is completed, low and high level features are extracted by convolution layers of the proposed architecture respectively. Afterwards, the system is trained by employing the labeled data by considering the predefined categories for the classification phase. Finally, the trained and optimized architecture is utilized for testing and validation processes. One of the critical contribution of this study is to propose a hybrid layer that involves three pretrained model instead of one for the transfer learning approach. The pretrained deep learning models are combined to generate a hybrid model for the proposed deep learning architecture, namely, VGG-16 (Simonyan and Zisserman 2014), GoogleNet (Szegedy et al. 2014), and Resnet (He et al. 2016). Seven distinct facial emotion categories are classified for the classification of the data. These categories are namely, “angry”, “disgust”, “fear”, “happiness”, “sadness”, “surprise” and “neutrality”.

As aforementioned, the architecture involves four main stages, which are the formation of the data set, the design of the deep neural network architecture, the training of the proposed deep neural network architecture and the evaluation of the proposed architecture. Overall, the structure of the proposed book chapter consists of the literature review, proposed framework, experimental results, and conclusion respectively.

LITERATURE REVIEW

In recent years, various different solutions have been proposed due to the importance of facial emotions in the design of human-computer interaction systems (Soylu, Guzel, and Askerzade 2018), and lots of progress has been attained in this research field.

Facial emotion recognition systems can be divided into two groups according to the image frame (static-based) or video frames (dynamic based) that they use as input data in the feature representations. First, in static-based facial emotion recognition systems, the feature representation is encoded only to spatial information in the current single image, and facial emotion recognition operations is performed using only the static facial feature obtained by extracting the handcrafted attributes. Secondly, in dynamic based facial emotion recognition systems, expression dynamics are obtained by using the spatial-temporal properties of facial expressions sequences and in these systems, the use of face physics based models is greatly reduced, and facial emotion recognition processes is performed by using the facial features obtained by providing end-to-end learning from direct input images within the pipeline (Ko 2018; Kim et al. 2017; Li and Deng 2018; Walecki et al. 2017).

Since the studied system is represented on real-time videos, it is included in the dynamic based facial emotion recognition systems category and the methodology of the studied system consists of four main steps. These are explained and can be seen by employing Figure 1 respectively. Firstly, input image frames are taken from real-time video. Secondly, face detection is performed on the video frames received. Afterwards, preprocessing operations is performed so as to reduce noise, enhance images for faster, and better training process. In addition, data augmentation and normalization processes are carried out in this phase respectively.

The following step involves the process of feature extraction, the training and classification operations of the deep neural networks. Finally, facial emotion recognition operation is performed by employing the trained architecture. There is a huge literature for the facial emotion and gesture recognition systems. This section only involves the corresponding literature employed at the design of the proposed architecture. Therefore, only the pretrained CNN architectures used by the designing of the hybrid system is employed. Those three pretrained architectures have been discussed correspondingly.

VGG-16 Network

VGG-16 is a pretrained convolutional neural network architecture that is trained on more than a million images from the ImageNet database (Simonyan and Zisserman 2014). This network aims to investigate the effect of convolutional network depth on accuracy in large-scale image recognition settings. The VGG-16 network shown in Figure 2 consists of 16 convolutional layers and from start to finish offers a homogeneous deep learning architecture with 3 x 3 convolutions filters and 2 x 2 pooling processes. In addition, there exists three fully connected layers in this architecture, as shown in the corresponding figure. Finally, a Softmax classifier is integrated into the architecture so as to classify data with high accuracy.

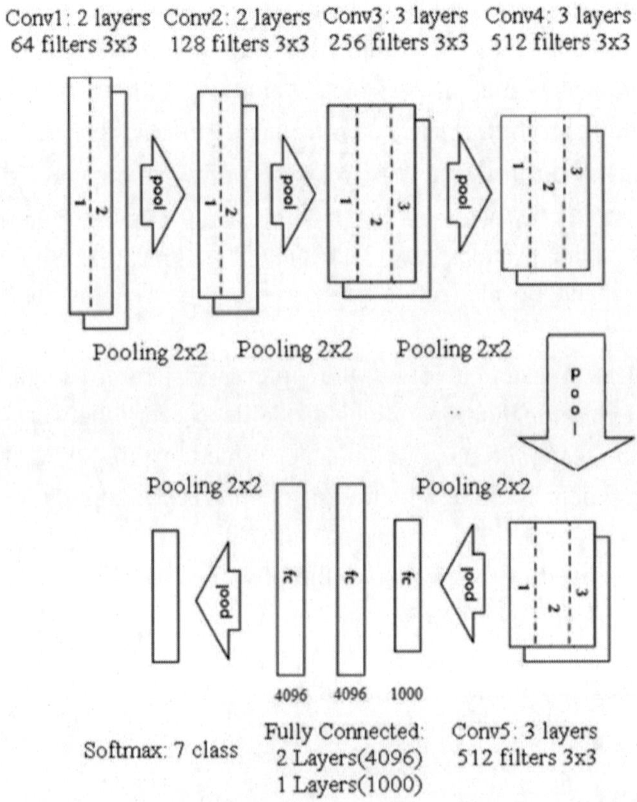

Figure 2. The basic architecture of VGG-16 network (Simonyan and Zisserman 2014).

The most significant superiority of the "VGG-16" network over the previous deep neural network architectures are the use of convolutional layers that doubles or triples and contains large number of filters that can strength the extraction of low and high level features respectively. Here, the values obtained from the convolution layers are converted into an attribute vector having "4096" neurons in the fully connected layers and those layers are connected with a Softmax layer allowing "1000" class for the classification phase. Here, Softmax layer is utilized to provide multi-class learning process in which a set of features can be related to one-of-K classes and its mathematical formulation is shown in equation (1). In this equation, $\sigma(z)_j$ represents softmax result, whereas z represents a vector of the inputs to the output layer and j indexes the output units. For this architecture, the depth value is increased while the height and width dimensions of the matrices decreases from input to output.

$$\sigma(z)_j = \frac{e^{z_j}}{\sum_{k=1}^{K} e^{z_k}} ,j = 1,2, \dots , K \tag{1}$$

The VGG-16 network is adapted to the proposed architecture for facial emotion recognition operations with a few small adjustments. According to which, first, the RGB image obtained from the video frames is given as the input to the network. Then, the preprocessing operations is performed by subtracting the calculated mean RGB value from each pixel. Afterwards, the image frame is convolved with 3 x 3 filters. During these processes, convolutional stride is fixed to one pixel. In addition, the spatial pooling operation is performed by relying on five max-pooling layers, which follows the convolution layers. The max-pooling process is performed on a 2 x 2 pixel window with two strides in here. Fourth, the stack of convolution layers follows three fully connected layers operations. The first two of the fully connected layers have 4096 channels and the third has 1000 channels. Finally, the facial emotion classification process is completed with inserting the features obtained from the fully connected layers into the Softmax layer and thus, facial emotion recognition operations are completed. In addition, rectified linear unit (RELU) function was used in all hidden layers within the VGG-16 network (Simonyan and Zisserman 2014). The RELU function is a non-linear activation function and is defined mathematically as in equation (2). With this equation, if the input is negative, the output is 0, and raw output otherwise.

$$f(x) = (0, x) \tag{2}$$

$$f(x) = \{\, 0, if\ x < 0\ x, if\ x \geq 0$$

GoogleNet Network

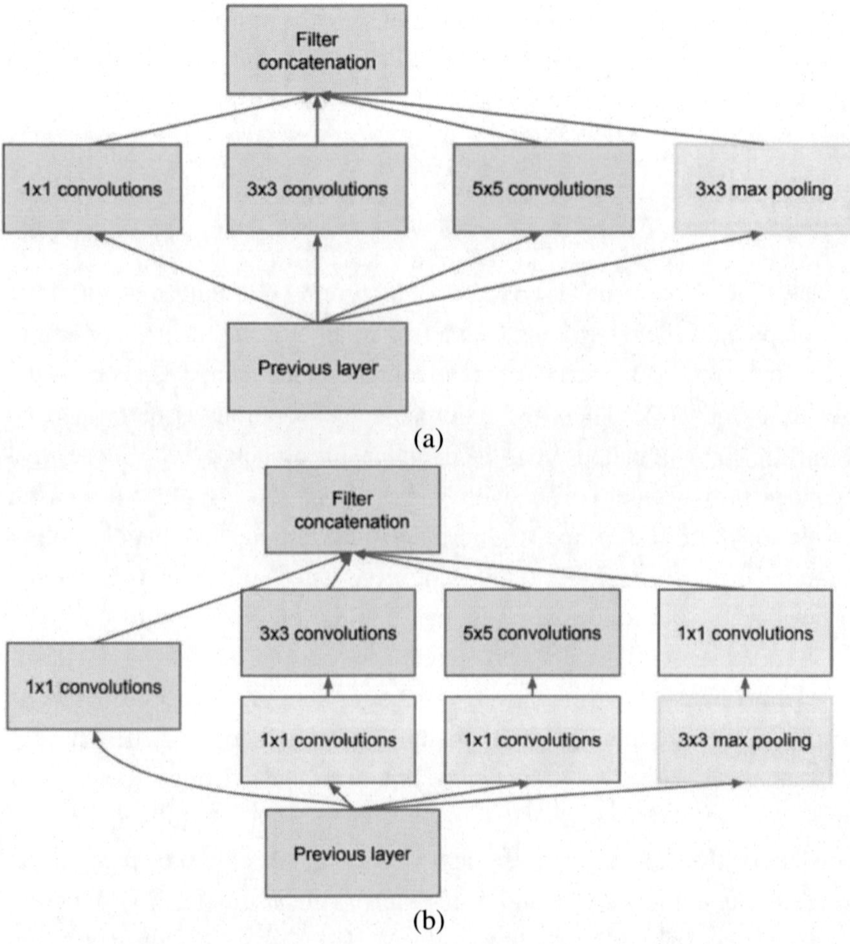

Figure 3. Inception module (a) Naive Inception module version (b) Inception module with dimension reductions (Szegedy et al. 2014).

Although the GoogleNet network has a complex architecture, it is the first architecture to use the concept of width within convolutional neural network models. The width item in the model is obtained by modules called inception. The expansion effect in the inception modules is obtained by performing “1 × 1”, “3 × 3”, “5 × 5” filters in the convolution layers and 3 × 3 maximum pooling operation in a parallel manner.

This module shown in Figure 3 (a) is called the Naive Inception Module. However, because the greatness of the output size and the number of parameters are grown too much due to parallel operations, the calculation complexity increases as it is expected. In order to solve this problem, the processing load is optimized employing a "1 x 1" convolution layer before each Naive Inception convolution layer and thus size reduction is performed successfully. The module, shown in Figure 3 (b), represents the first version of the inception Modules.

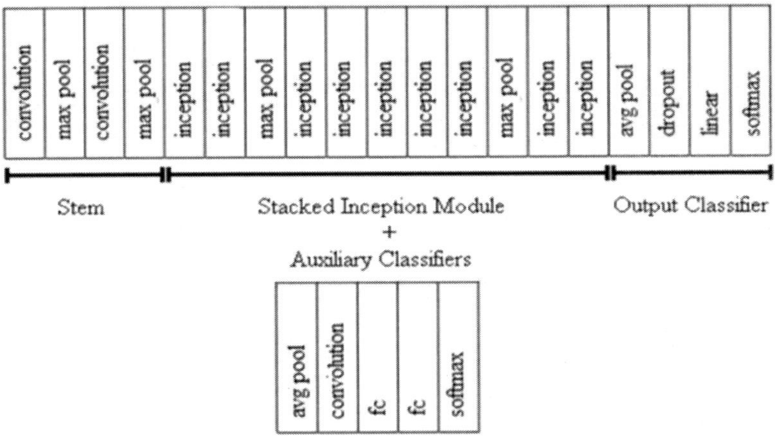

Figure 4. The basic architecture of GoogleNet network (Szegedy et al. 2014).

GoogleNet network architecture consists of a 22-layer architecture and has a depth of 27 layers if pooling layers are included. The architecture shown in Figure 4 consists of four main sections, namely stem, stacked inception module, auxiliary classifiers and output classifier. This architecture has 9 inception modules and 144 layers such as convolution, maxpooling, softmaxlayer, fullconnected layer, relulayer, input layer and output layer. This architecture also exploits "1 x 1", "3 x 3" and "5 x 5" size filters in its convolution layers, whereas "3 x 3" filters are used in pooling operations, and RELU (rectified linear unit) activation function are used for activation operations (Szegedy et al. 2014). For the proposed architecture, the GoogLeNet network is adapted as it has been performed in VGG-16 network. According to which several minor adjustments within the deep learning architecture mentioned above and thus, the facial emotion recognition operations was achieved.

Microsoft ResNet-152 Network

The Restnet-152 network work provides the residual learning framework to facilitate the training of deeper network networks that are problematic to train. According to which, a deep neural network architecture was designed to learn the residual functions by referring to the input layers rather than learning the unreferenced functions. The architecture of the Resnet-152 network is based on the VGG-16 network mentioned in Section 2.1. Figure 5 displays the basic architecture of this network that the convolution layer has "3 x 3" and "1 x 1" filters, followed by two pooling operations. Here, the downsampling process is completed by employing convolutional layers having 2 strides. The Resnet-152 network also possesses an average pooling layer, a 1000-way fully connected layer, and a softmax classifier. Essentially, the weighted layer numbers used to design this network is 152, shown in Figure 5.

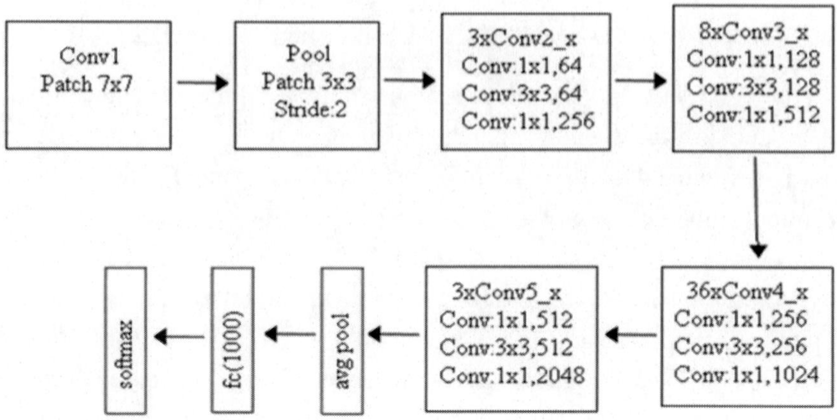

Figure 5. The basic architecture of Resnet 152 network (He et al. 2016).

The depth of a network is an important parameter in deep neural networks. According to the theory, the more the layer number increases, the more the representational capacity of the network is obtained. This approach essentially tends to create a better network architecture. However, this approach does not apply in practice. The reason for the lack of performance here is the vanishing (exploding gradients) and the difficulty of optimization problems. The problem of vanishing exploding gradients shows that the values of neurons without strong activations during the training of neural networks dissolve between the long layer chains and cannot contribute to the training phase of the network, whereas the optimization problem represents the increase at overall parameter numbers due to the formation of parameters originating from different structures. This prevents to estimate optimum parameter for the training phase in a short manner. However, one of the most important contribution of the ResNet network, compared with its counterparts, is that the network can prevent the negative situations that may arise due to the above mentioned problems while increasing the depth. Briefly, ResNet network, provides a depth architecture which can be trained faster and successfully. In order to solve the aforementioned problems, the residual block diagram is integrated into the Resnet architecture, as shown in Figure 6, and its equations are represented in equations (3) and (4) (He et al. 2016). While mapping the parts of the deep neural network architectures from the input to the exit with a nonlinear $H(x)$ function (shown in Figure 6a), it is mapped with another nonlinear function defined as $F(x) = H(x)\text{-}xH(x)$ instead of $H(x)$ in the structure shown in Figure 6b. Furthermore, a shortcut connections are made to the output from the input and the x input value is added arithmetically to the function $F(x)$. Then, the function $F(x) + x$ is passed through the ReLU activation function. Thus, by adding the input value to the end of the second layer, it is ensured that the values in the past layers are more strongly transmitted to the subsequent layers. While x and y represent the input and output vectors of the studied layers, and the $F(x,\{Wi\})$ function represents learning residual mapping for Equation 1, Equation (2), on the other hand, represents the second formulation to be used when dimensions of x and F are not equal. Besides, W_s represents the square matrix used to perform a linear projection with shortcut connections for this equation (He et al. 2016). For the proposed architecture, the Resnet-152 network was adapted for the facial emotion recognition operations.

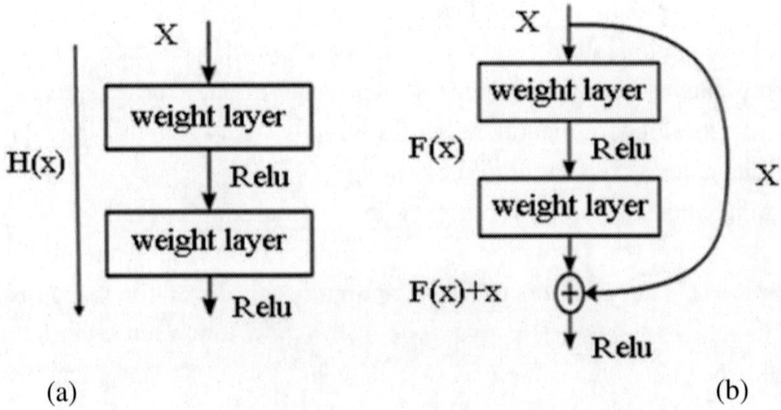

Figure 6. Residual block diagram (a) Basic Deep Neural Network Block (b) Deep Neural Network Block with Residual Block Connection (He et al. 2016).

$$y = F(x, \{W_i\}) + x. \quad (3)$$

$$y = F(x, \{W_i\}) + W_s x. \quad (4)$$

THE PROPOSED ARCHITECTURE

The main aim of this study is to propose a framework for the facial emotion recognition operations in an optimized way by designing a hybrid deep neural network architecture. The general architecture of the proposed framework is also illustrated in Figure 7.

The design concept of framework involves four main stages. These are namely, the creation of the data set, the design of the deep neural network architecture, the training and optimization phase and finally the validation of the trained deep neural network. Accordingly, first, the creation of the data set was accomplished. The data set consists of several comprehensive datasets such as The Japanese Female Facial Expression (JAFFE) (Lyons et al. 1998), The Extended Cohn-Kanade (CK +) (Lucey et al. 2010) and also personal photo collection of authors and their colleagues. CK+ dataset consists of a group of images representing 593 different emotions collected from 123 different subjects whereas the JAFFE dataset consists of a group of images representing 593 different emotions collected from 123 different subjects and our personal data set consists of a group of images that express 458 different emotions collected from 12 different subjects including data augmentation variants. As it has been previously declared, the proposed framework classifies facial emotions into seven categories, namely angry, disgust, fear, happiness, sadness, surprise and neutrality. Consequently, the pictures in the dataset are arranged according to these categories. Sample images of each facial emotion from the predefined dataset is shown in Figure 8 and the distribution of images in this customized dataset is also illustrated in Figure 9.

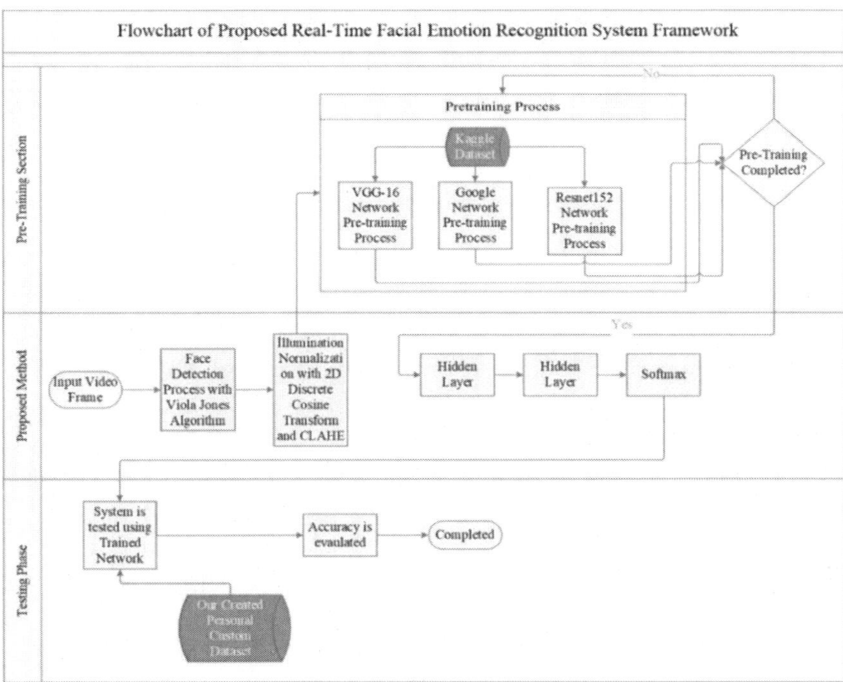

Figure 7. Flowchart representation of the proposed deep learning architecture.

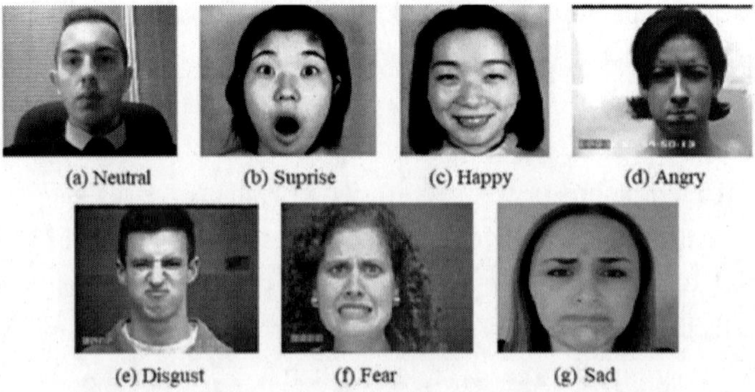

Figure 8. Sample images of each facial emotion dataset.

Emotion Classes	Sample Image Numbers	Percentage Rate(%)
Angry	176	14(%)
Disgust	189	15(%)
Fear	168	13(%)
Happy	214	17(%)
Sad	164	13(%)
Suprise	227	18(%)
Neutral	126	10(%)
Total	1264	100(%)

Figure 9. Sample image numbers of facial emotion classes in the created personal custom data set.

Once the dataset is obtained, the design operation of the deep neural network architecture is initialized. Here, two critical preprocessing steps have been achieved for the design of deep neural network architectures. The first of those step is to estimate an appropriate deep learning architecture to be used for the recognition process of facial emotions. Preliminary experiments reveal that a hybrid module may generate better overall accuracy. Hence, a hybrid module is generated by employing previously trained architectures. The hybrid module involves, VGG-16, Resnet152 and GoogleNet architectures. The details of these deep neural network architectures are described in Section 2. Afterwards it is aimed to explore the transfer learning approaches based on selected deep learning architectures in order to design the proposed deep neural network architecture. It should be noted that it may not be possible to obtain millions of datasets or process a large set of data in each application to done classification for use in a variety of processes, or in each model to be designed, and in this circumstance it would be more appropriate to utilize a previously trained model. The structure representing this situation is called transfer learning (Kaya, Keçeli, and Can 2018). In the transfer learning process, the weight values in the structure of the appropriately trained network for another dataset are taken and the training is carried out in a shorter time by adapting these weight values into a new dataset. This is basically is to prevent reinventing the wheel. As a result, due to the circumstances previously mentioned the transfer learning approach is adapted into the proposed architecture. In the proposed architecture, the low and high level features of images obtained from different architectures are combined. Figure 7 illustrates the structure of the proposed deep neural network architecture. The pre-trained networks shown in the training section of Figure 7 serves as feature extractors for general image features, and the last three layers represents completely fully-connected layers for the learning process and the softmax classifier for the classification operation. The details of the attributes produced by the selected pretrained deep learning architectures are summarized as follows:

VGG-16 Network: On each picture frame, a 2048-dimensional feature is extracted from the last fully connected layer as shown in Figure 2.

GoogleNet Network: On each picture frame, a 2048-dimensional feature is extracted from the last fully connected layer as shown in Figure 4.

Resnet152 Network: On each picture frame, a 2048-dimensional feature is extracted from the last fully connected layer as shown in Figure 5.

In the proposed architecture, firstly, the selected deep learning architectures were trained with Kaggle data set ('The Kaggle Dataset from the Facial Expression Recognition Challenge' n.d.) consisting of 35887 images and thus pre-training operation was performed. Secondly, attributes acquired from three deep learning architectures are combined to form a 6144 dimensional feature vector. Thirdly, the combined feature vector is passed into the fully connected layers and the Softmax layer for achieving the learning and classification phases respectively. This layer aims to increase the learning ability of the proposed network and to facilitate adapting the general features obtained by deep neural networks to facial image dataset. Finally, experimental analysis and validation is performed by employing the customized dataset as input to the trained model.

In order to improve the quality of learning process during the training phase, faces are detected in real time and aligned for the training section. One of the most popular face detection algorithm developed by Viola Jones is used in this study to extract the faces from an overall image (Viola and Jones 2004). This algorithm consists of three basic sections: the integral image, the Adaboost learning algorithm and the Haar-cascade classifier. Here, the integral image provides a very fast calculation of the rectangular properties and its mathematical formulation is shown in equation (5).

$$\begin{aligned} &ii(x, y) = \sum i(x', y') \\ &x' \leq x \\ &y' \leq y \end{aligned} \tag{5}$$

where ii (x, y) represents the integral image, (x, y) represents the point where the image is located and i (x, y) represents the image given as input

The Adaboost algorithm allows both the selection of the best haar-like attributes and the classification of the attributes. Haar cascade classifier algorithm is used to sense a specific object within a digital image or video frame. The decision-making mechanism of the Haar-cascade classifier algorithm is shown in equations (6) and (7).

$$C_m = \begin{cases} 1, \sum_{i=0}^{l_m-1} Fm, i > \theta_m \\ \quad 0, otherwise \end{cases} \tag{6}$$

Here, $f_{m,i}$ is the weighted sum of 2-dimensional integrals and i. is the decision threshold for the feature extractor.

$$F_{m,i} \begin{cases} \alpha_{m,i}, if\ f_{m,i} > t_{m,i} \\ \quad \beta_{m,i}, otherwise \end{cases} \tag{7}$$

where α_ (m, i) and β_ (m, i) represents constant values related with the i_{th} feature extractor, whereas θ_m represents the threshold of the m_{th} classifier. Within the revealed face detection algorithm, a rectangular window having the dimensions of 24 x 24 pixels, similar to the Haar wavelet patterns, are passed across the entire image to detect face images, which is called as the sliding window approach. Preprocessing operations are required on the image frames obtained as a result of the first stage. Here, the clipped images are first converted to the grayscale format and then the illumination normalization process is performed by applying the 2D Discrete Cosine Transform (W. Chen, Er, and Wu 2006) and Adaptive Histogram Equalization (CLAHE) together (Thamizharasi and Jayasudha, n.d.). With CLAHE method (Zuiderveld 1994) grayscale images are divided into 8 x 8 regions and then histogram equalization was applied to arrange the contrast of each region respectively. The obtained images were then resized to 224 x 224 pixels and converted to 3 channels. Finally, synthetic image data were generated using the data augmentation techniques from the obtained result image frames in the second stage. The aim here is to increase the number of data we have and to provide better training of our recommended deep neural network. The data augmentation operations are mainly employed such as rotation shifting, skew, scaling, noise, and color jittering.

EXPERIMENTAL RESULTS

This section details the implementation details and experimental results of the evaluation. Experiments were conducted with an Intel Core 2 Quad machine running at 2.8 GHz, with eight Gbytes of RAM memory in a Linux operating system.

In order to implement the proposed architecture, python programming language and its libraries (OpenCV, Numpy, Tensorflow, Dlib, Scikit-learn, Orange and so on) are employed. The training and testing procedures are carried out respectively. The dataset is shown in Figure 9 having 1294 images is employed during the training and test procedures. 70% of the available data is allocated for the training phase whereas 20% of them is allocated for the test procedure and 10% is used for the validation process respectively. The training process of the network architecture is carried out on a computer previously defined, without GPU support, for about 40 hours and the training was stopped at 200 epochs. For the evaluation procedure a simple but efficient evaluation metric is utilized. The evaluation metric value is the first step to evaluate the performance of the proposed model and thus the comparison process was carried out by comparing three deep neural networks with the proposed hybrid model. For the evaluation process, the accuracy parameter used in this section. The accuracy value gives the accuracy ratio of the models by measuring the ratio between the evaluated set and the correct results and is calculated by the ratio of the number of correctly classified samples to the total number of samples. In the study, this metric was calculated using the equation (8) shown below.

$$Accuracy = \frac{TP+TN}{TP+FP+FN+TN} \quad (8)$$

Table 1. Hyperparameter configurations for JAFFE, CK+, personal datasets

Parameter	JAFFE Dataset	CK+ Dataset	Personal Dataset
Epoch	35	90	75
Regularization	0.001	0.001	0.001
Batch Size	32	64	64
Learning rate 0-500 iteration 500-1000 iteration 1000+ iteration	 0.01 0.005 0.002	 0.01 0.005 0.002	 0.01 0.005 0.002
Momentum	0.9	0.9	0.9
Dropout	0.5	0.5	0.5
Weight Initialization	Xavier Initialization	Xavier Initialization	Xavier Initialization
Optimization Algorithm	SGD	SGD	SGD
Loss Function	Categorical Cross Entropy	Categorical Cross Entropy	Categorical Cross Entropy

Here, TP refers true positive, FP is false positive, whereas TN is for true negative and TP is for true positive.

While, Table 1 indicates the configuration parameters defined for each experimental datasets, Table 2 shows the accuracy metric values of the proposed network and other VGG-16, GoogleNet and ResNet152 deep neural network models on the individual CK +, JAFFE and customized datasets. According to corresponding table, it has been noted that the proposed method mostly outperforms those contemporary architectures as it is expected. In addition, the performances of the other three deep neural networks vary considerably with respect to the three sets of data, while the performance of our network shows similar performance results on those three dataset. These mentioned situations are showed that our network is more robust and superior performance than the other three deep neural networks. In order to better understand and evaluate the performance of our network model, secondly, in addition to the mentioned metrics, each emotion category was examined together with the confusion matrices. Figure 10 presents the confusion matrices obtained by the customized dataset for the proposed network and other deep neural network models. In this figure, the confusion matrices obtained on the created personal custom dataset for the studied categories of VGG-16, GoogleNet, ResNet152 and proposed network models are shown, respectively, from left to right and top to bottom. It is observed that confusion matrix of the proposed method shown in Figure 10.d outperforms other three network models in all the emotional situations except sadness. In addition, all network models can easily distinguish happiness, surprise, and neutral emotions. The Resnet-152 network model, shown in Figure 10.c, provides a better sense of sadness emotion result than other network models. In order to evaluate the performance of the proposed model in our study, finally, comparison operation with the state-of-the-art results directly cited from their publications are carried out. Table 3 illustrates the accuracy values obtained on the CK + data set for the proposed network model and other state-of-the-art studies. It is noted that the performance of the proposed architecture is more successful than other state-of-the-art algorithms by providing a higher accuracy value. An example output obtained from the framework is given in Figure 11.

Table 2. The obtained accuracy metric values of the the proposed architecture and other contemporary deep neural network models on the individual JAFFE (a), CK+ (b) and customized data set (c)

Method	Accuracy(%)
VGG-16 Network [12]	83,9
GoogleNet Network [13]	79,7
Resnet152 Network [14]	84,6
Proposed Network	**81,5**

(a)

Method	Accuracy(%)
VGG-16 Network [12]	82,8
GoogleNet Network [13]	77,2
Resnet152 Network [14]	85,7
Proposed Network	**92,5**

(b)

Method	Accuracy(%)
VGG-16 Network [12]	86,5
GoogleNet Network [13]	89,2
Resnet152 Network [14]	89,6
Proposed Network	**93,6**

(c)

Table 3. Facial emotion recognition accuracy table on CK+ database for proposed network and other state-of-the-art algorithms results

Method	Accuracy (%)
3DCNN [27]	85,9
ITBN [28]	86,3
STM [29]	91,13
LOMo [30]	92,0
AUDN [31]	92,05
3DCNN-DAP [27]	92,4
Proposed Network	**92,5**

	Angry	Disgust	Fear	Happy	Sad	Suprise	Neutral
Angry	82,3	5,8	2,6	1,9	3,8	1,9	1,7
Disgust	7,2	76,0	4,4	3,8	4,4	1,5	2,7
Fear	3,9	2,1	82,2	2,7	6,8	1,5	0,8
Happy	0,0	0,0	2,4	94,1	1,9	0,0	1,6
Sad	3,9	1,9	2,7	2,1	81,9	1,7	5,8
Suprise	0,0	0,3	1,2	0,0	2,1	92,0	4,4
Neutral	0,1	0,0	1,5	0,9	0,6	0,2	96,7

(a)

	Angry	Disgust	Fear	Happy	Sad	Suprise	Neutral
Angry	86,8	4,4	1,8	1,7	2,6	0,8	1,9
Disgust	5,8	81,7	4,3	2,6	1,8	0,9	2,9
Fear	5,7	2,2	84,6	1,1	3,3	1,2	1,9
Happy	0,0	0,0	1,6	96,3	1,7	0,0	0,4
Sad	3,3	0,9	2,9	2,2	84,9	1,5	4,3
Suprise	0,0	0,0	0,7	0,0	1,8	94,0	3,5
Neutral	0,0	0,0	1,7	0,3	1,6	0,6	95,8

(b)

	Angry	Disgust	Fear	Happy	Sad	Suprise	Neutral
Angry	88,5	4,1	1,1	1,2	3,1	0,9	1,1
Disgust	5,5	80,8	4,9	3,7	2,2	1,1	1,8
Fear	4,9	0,7	83,2	1,8	5,7	1,1	2,6
Happy	0,0	0,0	2,1	95,2	2,7	0,0	0,0
Sad	2,5	0,4	2,6	1,4	87,9	0,5	4,7
Suprise	0,0	0,0	0,4	0,0	2,3	94,4	2,9
Neutral	0,2	0,0	1,3	0,1	1,2	0,1	97,1

(c)

	Angry	Disgust	Fear	Happy	Sad	Suprise	Neutral
Angry	93,9	3,2	0,7	0,4	1,4	0,1	0,3
Disgust	3,2	89,4	0,1	1,2	3,7	1,0	1,4
Fear	1,7	1,9	87,2	3,8	2,9	1,3	1,2
Happy	0,0	0,0	0,0	98,6	1,4	0,0	0,0
Sad	2,6	0,1	1,9	0,2	87,4	1,0	6,8
Suprise	0,0	0,0	0,8	0,0	0,0	99,1	0,1
Neutral	0,1	0,0	0,0	0,0	0,1	0,1	99,7

(d)

Figure 10. The confusion matrix representations obtained on the customized dataset for the studied emotion categories of (a) VGG-16 Network (b) GoogleNet Network (c) ResNet152 Network (d) Proposed Network.

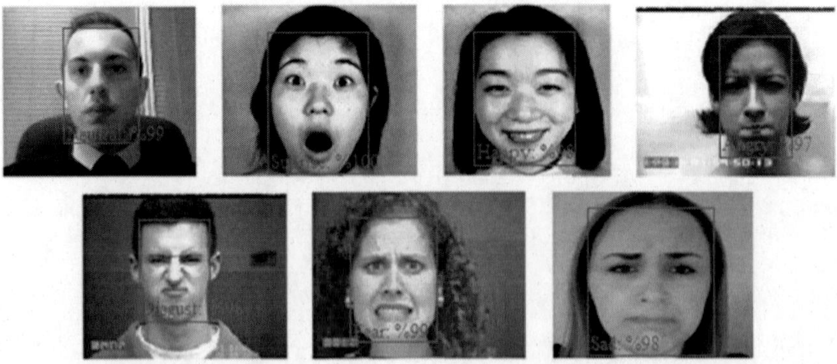

Figure 11. The facial emotion classification results obtained from the proposed framework.

Conclusion

This study proposes a new deep learning based architecture for the recognition of facial emotions and gestures. The architecture is basically designed based on pretrained networks using the transfer learning methodology. The main contribution of the proposed method is to propose a hybrid model combining three well-known pretrained networks in an optimized manner. The performance of the proposed architecture is evaluated by employing comprehensive datasets including three popular datasets and a customized dataset generated for this study. Once the proposed hybrid model is compared with the individual models, the proposed model almost outperforms than all popular individual models for all scenarios.

Besides these experiments are also compared with the state of the art studies published in various comprehensive academic papers. Those results also reveal and validate the superiority of the proposed method over the state-of-the-art methods. This hybrid approach also encourages authors to generate hybrid models from pretrained models to overcome more complicated computer vision and related tasks.

REFERENCES

Bartneck, Christoph, and Michael J. Lyons. 2007. 'HCI and the Face: Towards an Art of the Soluble.' In *International Conference on Human-Computer Interaction*, 20–29. Springer.

Chen, Chien-Hsu, I.-Jui Lee, and Ling-Yi Lin. 2015. 'Augmented Reality-Based Self-Facial Modeling to Promote the Emotional Expression and Social Skills of Adolescents with Autism Spectrum Disorders.' *Research in Developmental Disabilities* 36: 396–403.

Chen, Weilong, Meng Joo Er, and Shiqian Wu. 2006. 'Illumination Compensation and Normalization for Robust Face Recognition Using Discrete Cosine Transform in Logarithm Domain.' *IEEE Transactions on Systems, Man, and Cybernetics, Part B (Cybernetics)* 36 (2): 458–466.

Dornaika, Fadi, and Bogdan Raducanu. 2007. 'Efficient Facial Expression Recognition for Human Robot Interaction.' In *International Work-Conference on Artificial Neural Networks*, 700–708. Springer.

Essa, Irfan A., and Alex Paul Pentland. 1997. 'Coding, Analysis, Interpretation, and Recognition of Facial Expressions.' *IEEE Transactions on Pattern Analysis and Machine Intelligence* 19 (7): 757–763.

Gao, Hua, Anil Yüce, and Jean-Philippe Thiran. 2014. 'Detecting Emotional Stress from Facial Expressions for Driving Safety.' In *2014 IEEE International Conference on Image Processing (ICIP)*, 5961–5965. IEEE.

Güzel, Mehmet Serdar. 2016. 'Performance Evaluation for Feature Extractors on Street View Images.' *The Imaging Science Journal* 64 (1): 26–33.

He, Kaiming, Xiangyu Zhang, Shaoqing Ren, and Jian Sun. 2016. 'Deep Residual Learning for Image Recognition.' In *Proceedings of the IEEE Conference on Computer Vision and Pattern Recognition*, 770–778.

Hickson, Steven, Nick Dufour, Avneesh Sud, Vivek Kwatra, and Irfan Essa. 2019. 'Eyemotion: Classifying Facial Expressions in VR Using

Eye-Tracking Cameras.' In *2019 IEEE Winter Conference on Applications of Computer Vision (WACV)*, 1626–1635. IEEE.

Kaulard, Kathrin, Douglas W. Cunningham, Heinrich H. Bülthoff, and Christian Wallraven. 2012. 'The MPI Facial Expression Database—a Validated Database of Emotional and Conversational Facial Expressions.' *PloS One* 7 (3): e32321.

Kaya, Aydın, Ali Seydi Keçeli, and Ahmet Burak Can. 2018. 'Examination of various classification strategies in classification of lung nodule characteristics' Journal of the Faculty of Engineering and Architecture of Gazi University, 2018 (18–1).Kim, Dae Hoe, Wissam J. Baddar, Jinhyeok Jang, and Yong Man Ro. 2017. 'Multi-Objective Based Spatio-Temporal Feature Representation Learning Robust to Expression Intensity Variations for Facial Expression Recognition.' *IEEE Transactions on Affective Computing* 10 (2): 223–236.

Ko, Byoung. 2018. 'A Brief Review of Facial Emotion Recognition Based on Visual Information.' *Sensors* 18 (2): 401.

Li, Shan, and Weihong Deng. 2018. 'Deep Facial Expression Recognition: A Survey.' *ArXiv Preprint ArXiv:1804.08348*.

Lucey, Patrick, Jeffrey F. Cohn, Takeo Kanade, Jason Saragih, Zara Ambadar, and Iain Matthews. 2010. 'The Extended Cohn-Kanade Dataset (Ck+): A Complete Dataset for Action Unit and Emotion-Specified Expression.' In *2010 IEEE Computer Society Conference on Computer Vision and Pattern Recognition-Workshops*, 94–101. IEEE.

Lyons, Michael J., Shigeru Akamatsu, Miyuki Kamachi, Jiro Gyoba, and Julien Budynek. 1998. 'The Japanese Female Facial Expression (JAFFE) Database.' In *Proceedings of Third International Conference on Automatic Face and Gesture Recognition*, 14–16.

Mehrabian, Albert. 2008. 'Communication without Words.' *Communication Theory*, 193–200.

Muhammad, Ghulam, Mansour Alsulaiman, Syed Umar Amin, Ahmed Ghoneim, and Mohammed F. Alhamid. 2017. 'A Facial-Expression Monitoring System for Improved Healthcare in Smart Cities.' *IEEE Access* 5: 10871–10881.

Mutlu, Begüm, Murat Hacıömeroğlu, Guzel Mehmet Serdar, Mehmet Dikmen, and Hayri Sever. 2014. 'Silhouette Extraction from Street View Images.' *International Journal of Advanced Robotic Systems* 11 (7): 114.

Simonyan, Karen, and Andrew Zisserman. 2014. 'Very Deep Convolutional Networks for Large-Scale Image Recognition.' *ArXiv: 1409.1556 [Cs]*, September. http://arxiv.org/abs/1409.1556.

Soylu, Busra Emek, Mehmet Serdar Guzel, and I. N. Askerzade. 2018. *'A New Deep Learning Based System for Facial Gesture Analysis.'*

Srivastava, Saumil. 2012. *'Real Time Facial Expression Recognition Using a Novel Method.'* ArXiv Preprint ArXiv:1206.3559.

Szegedy, Christian, Wei Liu, Yangqing Jia, Pierre Sermanet, Scott Reed, Dragomir Anguelov, Dumitru Erhan, Vincent Vanhoucke, and Andrew Rabinovich. 2014. *'Going Deeper with Convolutions.'* ArXiv: 1409.4842 [Cs], September. http://arxiv.org/abs/1409.4842.

Thamizharasi, A., and J. S. Jayasudha. n.d. *'An illumination invariant face recognition using 2D discrete cosine transform and clahe.'*

'The Kaggle Dataset from the Facial Expression Recognition Challenge.' n.d. Accessed 1 May 2019. https://www.kaggle.com/c/challenges-in-representation-learning-facial-expression-recognitio%20n-challenge/data.

Viola, Paul, and Michael J. Jones. 2004. 'Robust Real-Time Face Detection.' *International Journal of Computer Vision* 57 (2): 137–154.

Walecki, Robert, Ognjen Rudovic, V. Pavlovic, B. Schuller, and Maja Pantic. 2017. 'Deep Structured Learning for Facial Expression Intensity Estimation.' *Image Vis. Comput* 259: 143–154.

Zuiderveld, Karel. 1994. 'Contrast Limited Adaptive Histogram Equalization.' In *Graphics Gems IV*, 474–485. Academic Press Professional, Inc.

In: Human-Computer Interaction
Editor: Turgut Ozseven
ISBN: 978-1-53616-495-4

Chapter 3

DESIGN OF A VERSATILE AND RAPID CALCULATOR FOR ONCOLOGICAL COMPUTATIONS

Çiğdem Dinçkal*
Department of Civil Engineering,
Çankaya University, Ankara, Turkey

ABSTRACT

In oncology, there exists many complicated and time-consuming computations for students, clinicians and medical researchers. Although many programs generated in Microsoft Excel have been studied many times, the improvements in design of programs are still an issue of concern. So design of calculator becomes essential for practical use. The objective of this study is to design a rapid and versatile calculator named as TGTSCalculator in Microsoft Excel for comprehensive computation of clonogenic cells amount when growth and irradiation for tumor occurred for all tumor growth models and the total amount of relative cells outliving the treatment schedule (in other words; survival fraction) after radiotherapy.

* Corresponding Author's Email: cdinckal@cankaya.edu.tr.

Microsoft Excel with Visual Basics for Applications (VBA) has been employed since it is a high versatile and extensively used special spreadsheet program and easily accessible and available for biomedical scientists and clinicians for a wide range of applications. TGTSCalculator is so transparent and practical that any kind of users requires only requires clicking its macro or CALCULATE button in spreadsheet to manage the all computations effectively and simultaneously.

Keywords: tumor growth models, treatment schedule, clonogenic survival fraction, Microsoft Excel, Visual Basics for Applications

INTRODUCTION

Cancer is still the primary reason of death in the world yet much not known about its mechanisms of detection and destruction (Kirschner and Panetta 1998). Furthermore, tumor population is much more instable, appearing with epigenetic, its genetic and micro-environmental heterogeneity a constantly thriving spectrum of tumor behaviors and cell expressions (Hahnfeldt et al. 1999). In this sense, tumor cell repopulation and tumor repopulation kinetics between treatments and also treatment schedules become significant.

Besides, there exists many cancer treatments and studies about treatment in literature such as anti-angiogenic therapy, imaging techniques, chemotherapies, radiotherapies which directly kill tumor cells or prevent their proliferation (Chudej, Wagner, and Pesch 2015; Dixit et al. 2012; Petrakis, Petraki, and Petrakis 2017; Yankeelov et al. 2010; Rejniak and Enderling 2013; McAneney and O'Rourke 2007; Sachs, Hlatky, and Hahnfeldt 2001; Enderling and AJ Chaplain 2014). It is important to select the optimum treatment type and its schedule for the related tumor repopulation kinetics. The laws for tumor cell repopulation involves four different growth mechanisms, named as, exponential model, restricted exponential model, Gompertz model and logistic model (McAneney and O'Rourke 2007; Wheldon and Kirk 1976; Wheldon, Kirk, and Orr 1977; Panetta and Adam 1995).

In this study, these four mechanisms were considered with corresponding radiation treatments and thereby treatment schedules. Treatments by radiotherapy, one of the crucial weapon of oncologist, were taken in consideration. Related treatment schedules observations involve clonogenic survival fraction with the single radiation dosage D and α/β ratio (an inverse measure of a tissue's sensitivity to fractionation), here α and β are radiobiological parameters (taken from advanced head, neck and prostate cancer as well as radiosensitive parameters in literature) for the tumor issue concerned for daily treatments (Joiner and Van der Kogel 2016; Thames et al. 1990; Carlson et al. 2004; King, DiPetrillo, and Wazer 2000). Standardized treatment protocols are also taken into account.

At this stage, long and tedious formulations are required to be computed for each tumor growth mechanisms and related survival fraction in treatment schedules. One of the main purposes of this study is to focus on computations with a user-friendly, rapid and high versatile macro called as TGTSCalculator created in Microsoft Excel with contribution of VBA. Although Microsoft Excel was employed in many studies in literature, neither of them used Microsoft Excel for this purpose. Instead, Microsoft Excel was employed for data management and analysis, paternity analysis, statistical analysis such as regression, Monte Carlo simulation in remote clinics, biology, biological systems, pharmacology, medicine in literature (Gomes, Passeri, and de Albergaria Barbosa 2006; Scribner Jr et al. 2001; Craighead et al. 2000; Achuthan, Grover, and MacFie 2006; Glazebrook et al. 2005; Brown 2000, 1999, 2006; Wu 2004; Wera 1998; Pfaffl et al. 2004; Geeraerd, Valdramidis, and Van Impe 2005; Dansirikul, Choi, and Duffull 2005; Messmer 2005; Aoki et al. 1997; Rocheta et al. 2007; Khan 2005; Anbazhagan 2003; Khan 2004; Schageman et al. 2002; Conway et al. 2002; Rathod et al. 2014; Khan 2006; Brown 2001, 2010; Lambert et al. 2012). Moreover, some studies in literature employed Microsoft Excel with VBA for data analysis and statistical analysis in biology and especially pharmacology (Zhang et al. 2010; Brocks 2015; Meineke 2000; Gilani, Moghadam, and Zakerhamidi 2011; Zou and Li 2013; Greensmith 2014; Sato et al. 1996).

Another purpose of the study is to perform all computations with comparative analyses and speed of the calculator in terms of seconds, simultaneously. So, there is no need to memorize long and troublesome formulations and spend most of the time for computations. The only thing for user is to enter the necessary inputs and obtain the accurate results.

METHODS

Computations in present study are based on radiotherapy (playing an important role in cancer treatment) and the mechanisms of how radiation results in DNA damage. For this reason, linear-quadratic (LQ) model was used to understand the radiation survival response for a solid tumor (McAneney and O'Rourke 2007; Sachs, Hlatky, and Hahnfeldt 2001). It remarked that for external beam radiotherapy, the clonogenic surviving fraction is (McAneney and O'Rourke 2007):

$$\sigma = exp(-\alpha D - \beta D^2) \tag{1}$$

Here D (Gy) is a single radiation dose and β (Gy^{-2}) and α (Gy^{-1}) are the radiobiological cell survival parameters dependent on tissue type within the volume being treated (McAneney and O'Rourke 2007). During radiotherapy, fractionation of treatment schedules occurs to permit to repair the normal tissue and recover from the irradiation. During these periods, surviving clonogenic cells of the tumor also repair and repopulate. This re-growth of specific tumor concerned can affect the outcome of a particular treatment schedule (McAneney and O'Rourke 2007).

During radiation treatment, the repopulation of a homogeneous tumor cell line is generally formulated as

$$N_1 = \sigma f(N_0, t), \tag{2a}$$

$$N_2 = \sigma f(N_1, t) = \sigma f(\sigma f(N_0, t), t), \tag{2b}$$

$$N_3 = \sigma f(N_2, t) = \sigma f(\sigma f(\sigma f(N_0, t), t), t), \tag{2c}$$

...

$$N_n = \sigma f(N_{n-1}, t) = \sigma f(\sigma f(\dots f(\sigma f(N_0, t), t), \dots, t), t), \tag{2d}$$

where N_i is the final number of clonogenic cells after the ith time interval, t, when the tumor has been irradiated and grown, σ is the clonogenic survival fraction of the tumor as obtained from eq. (1) and the function $f(N_i, t)$ denotes the growth mechanisms called as; exponential model, restricted exponential model, Gompertz model and logistic model (McAneney and O'Rourke 2007).

Growth Mechanisms

In the following subsections, the subscripts *e, r, g* and *l* are employed to characterise the exponential, restricted exponential, Gompertz and logistic models, respectively.

Exponential Model

Exponential model is defined by the following function:

$$N(t) = g_e N \tag{3}$$

where $g_e = \ln 2/t_2$ is the constant growth rate obtained by relating it to the tumor doubling time t_2 and N_0 is the initial number of cells at $t = 0$ from which growth is taken into account to evolve for a period of time t (McAneney and O'Rourke 2007).

Restricted Exponential Model

An extended version of the exponential growth model is restricted exponential growth (Panetta and Adam 1995). The difference of this model from exponential growth is due to the decrease in clonogenic cell rate as the tumor growths. This model is specified by the following function:

$$N(t) = K + (N_0 - K)\exp(-\mathrm{g_r}t) \quad (4)$$

where $\mathrm{g}_r = \mathrm{g}_e[N_0/(K - N_0)]$ is the growth rate of clonogenic cells in terms of exponential model growth rate, *K* is the carrying capacity or limiting size the tumor would reach without treatment or death (McAneney and O'Rourke 2007).

Gompertz Model

In Gompertz model, *N*, function of time takes the following form:

$$N(t) = K\exp\left[\ln(N_0/K)\exp\left(-\mathrm{g_g}t\right)\right] \quad (5)$$

where $\mathrm{g_g} = \mathrm{g}_e/\ln(K/N_0)$ is the growth rate of clonogenic cells (McAneney and O'Rourke 2007).

Logistic Model

This model is one of the form of self-limiting growth and described by the following function:

$$N(t) = \frac{N_0 K}{N_0 + (K - N_0)\exp(-\mathrm{g_l}t)} \quad (6)$$

where the growth rate is found as $\mathrm{g}_l = \mathrm{g}_e/[1 - (N_0/K)]$ (McAneney and O'Rourke 2007).

Treatment Schedules

These schedules can vary in terms of survival fractions according to the tumor growth mechanisms explained briefly in the following subsections.

Exponential Model

According to exponential growth, survival fraction after *n* uniform scheduled treatments at time intervals *t* can be computed by the following formula (McAneney and O'Rourke 2007):

$$\Sigma_{n,e} \exp[-n(\alpha D + \beta D^2)]\exp(n g_e t) = (\sigma/a_e)^n \tag{7}$$

where $\Sigma_{n,e} = N_n/N_0$ is the relative number of cells surviving fraction (or surviving the treatment schedule) and $a_e = \exp(-g_e t)$ [11].

Restricted Exponential Model

Due to restricted exponential growth, final survival fraction after *n* treatments at time intervals *t* obtained as follows:

$$\Sigma_{n,r} = (\sigma a_r)^n + \frac{K\sigma(1-a_r)(1-(\sigma a_r)^n)}{N_0(1-\sigma a_r)} \tag{8}$$

where $a_r = \exp(-g_r t)$ [8].

Gompertz Model

To the Gompertz growth, final survival fraction at the end of *n* treatments at time intervals *t* found as follows:

$$\Sigma_{n,g} = (K/N_0) + \exp\left(\frac{(1-a_g^n)}{(1-a_g)}\ln\sigma + a_g^n \ln(N_0/K)\right) \tag{9}$$

where $a_g = \exp(-g_g t)$ (McAneney and O'Rourke 2007).

Logistic Model

For logistic growth, surviving fraction after *n* treatments can be computed by use of following formula:

$$\Sigma_{n,l} = \sigma\left[\frac{a_l^n}{\sigma^{n-1}} + \frac{N_0(1-a_l)(1-(a_l/\sigma)^n)}{K(1-(a_l/\sigma))}\right]^{-1} \tag{10}$$

where $a_l = \exp(-g_l t)$ [8].

The growth from each model has been suitably scaled so that the same initial conditions exist.

Design of TGTSCalculator

Design of TGTSCalculator incorporates three stages described in the following subsections.

Description of TGTSCalculator

Using this calculator runs a macro named TGTSCalculator which performs all oncological calculations at once completely. The general procedure is consisting of some phases. These phases are general and applicable to any type of tumor growth model with different parameters and various treatment schedules. The first phase is to generate a user interface input form named UserForm1 to make users enter the inputs; initial number of cells, carrying capacity, doubling time (t_2), α, β, *D*, *n* (repetitions), time (period). For any type of problem, a typical UserForm1 is displayed in Figure 1. The second phase is to form a new tab called as TGTSCalculator with CALCULATE button including computations of each formulations given eq. (1) to eq. (10). This is shown in Figure 2.

Implementation of TGTSCalculator

Figure 2 illustrates the TGTSCalculator tab with CALCULATE button (assigned TGTSCalculator). By clicking this button, user is able to run the macro conveniently, and User input form in spreadsheet.

Sample Run of the Macro

After entering all inputs, user only needs to click Run button, then the screen takes the form in Figure3. Total elapsed time for all computations is also displayed in the message box in Figure 3.

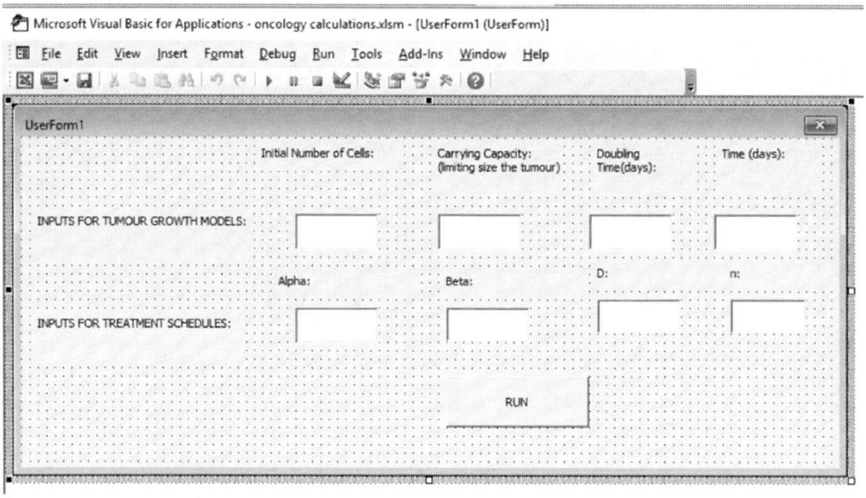

Figure 1. The standard UserForm1.

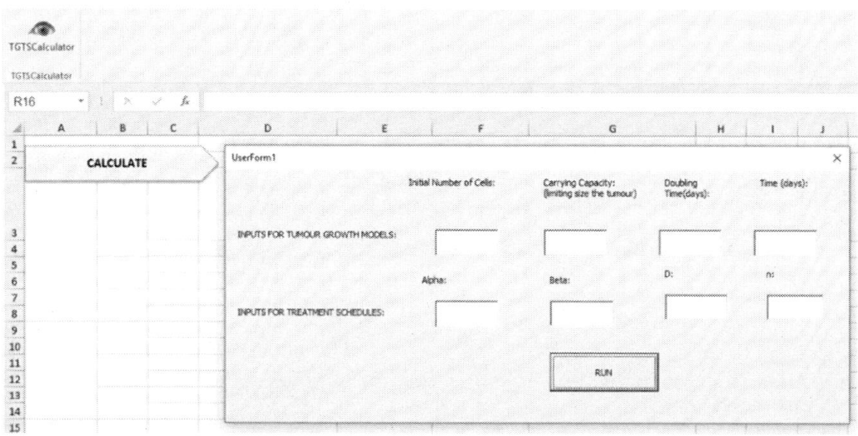

Figure 2. The standard TGTSCalculator with CALCULATE button and Unfilled UserForm1 for the example.

Closing the window of UserForm1 leads to Figure 4. Figure 4 (a) presents the computational results for final number of clonogenic cells in given period of time for each tumor growth model and Figure 4 (b) gives the clonogenic survival fraction for each growth mechanism.

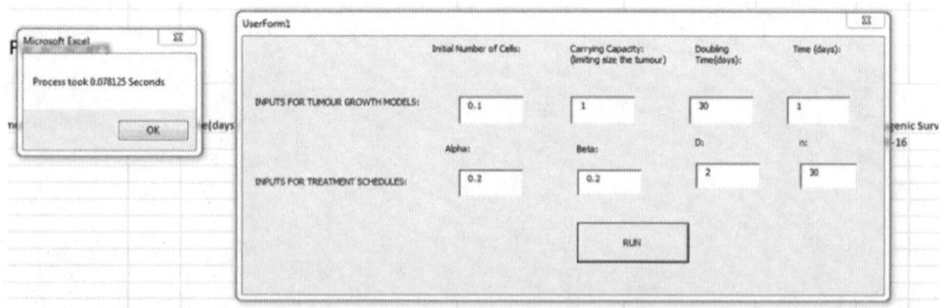

Figure 3. Filled UserForm1 for the example and total elapsed time for the computations.

	Tumour Growth Models					Parameters		
CALCULATE	Time (days)	Exponential Model	Restricted Exponential Model	Gompertz Model	Logistic Model	Alpha/Beta	Tumour Doubling Time	No/K
	1	0.102337389	0.102307527	0.102325566	0.102334333	1	30	0.1

(a)

Clonogenic Survival Fraction

Time (days)	Clonogenic Survival Fraction for Exponential Model	Clonogenic Survival Fraction for Restricted Exponential Model	Clonogenic Survival Fraction for Gompertz Model	Clonogenic Survival Fraction for Logistic Model
1	4.63905E-16	2.14949E-16	4.91311E-14	4.99166E-16

(b)

Figure 4. (a) Computation results for each tumor growth model (b) Computation results for clonogenic survival fraction of each growth model.

RESULTS

In this study, the role of different nonlinear growth laws, which define non-constant growth of a tumor are taken into account along with the impact that they will have on the output given various survival fractions.

The models analyzed involve the restricted exponential (Panetta and Adam 1995), the logistic and Gompertz models, and are compared with the assumed constant growth kinetics given by the exponential model with numerical example.

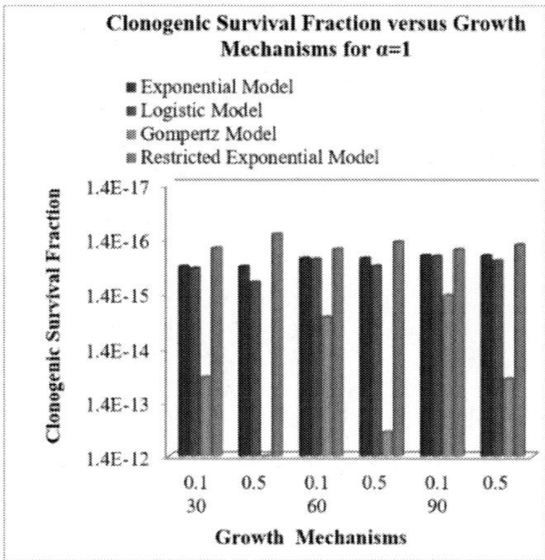

(a)

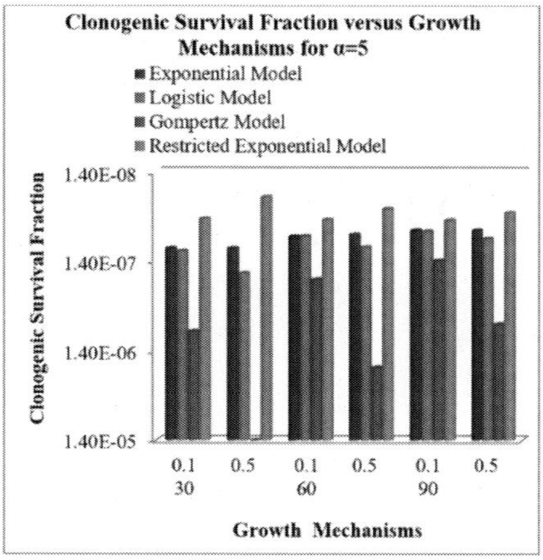

(b)

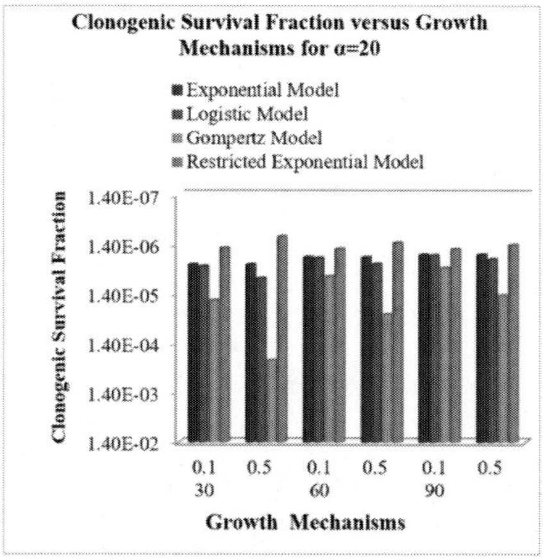

(c)

Figure 5. Computation results for clonogenic survival fraction for each growth mechanism when (a) $\alpha = 1$ Gy^{-1} (b) $\alpha = 5$ Gy^{-1} (c) $\alpha = 20$ Gy^{-1}.

For numerical example case, four tumor growth models are considered. The numerical data are given as $\alpha = 0.2$ Gy^{-1} $\beta = 0.2$ Gy^{-2} (takes also the values $\beta = 0.04$ Gy^{-2} and $\beta = 0.01$ Gy^{-2}), tumor doubling time (t_2) = 60 days (takes also the values 30 and 90 days), $D = 2$ Gy and $N_0 = N_0/K = 0.1$

(takes also the value N_0 = 0.5), with n = 30 repetitions (McAneney and O'Rourke 2007).

Entering these numerical data and running calculator bring forth Figure 5, 6 and 7. In Figure 5, the effect of varied ratio of α/β on clonogenic survival fraction for different growth mechanisms are illustrated. t_2 changed as 30 60, 90 days and N_0 changed as 0.1 and 0.5 on the horizontal axis while vertical axis gives the clonogenic survival fraction.

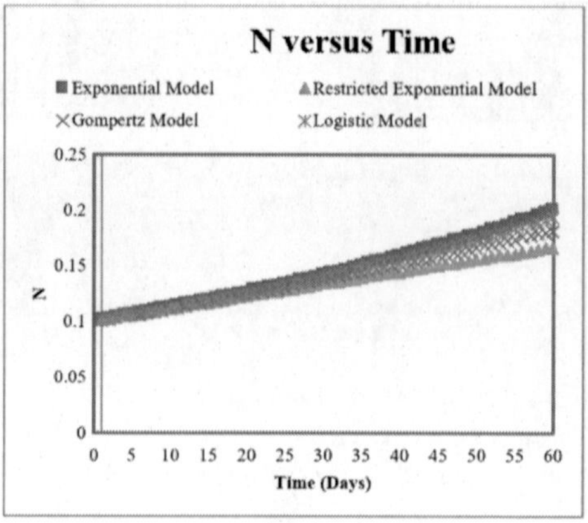

(a)

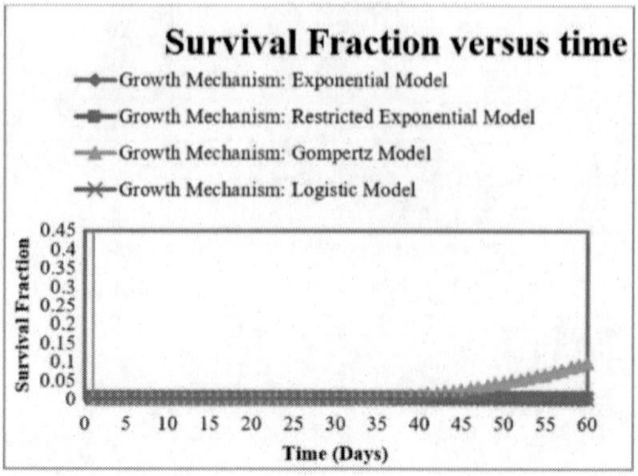

(b)

Figure 6. (a) Final number of clonogenic cells in 60 days periods for each growth mechanism (b) Clonogenic survival fraction in 60 days periods for each growth mechanism.

Low α/β values indicate that, tissues have a high repair capacity, on the contrary, high α/β leads to a low repair capacity. So, Fig. 5 indicates that Gompertz model reveals poorest prognosis among the other growth models for each values of α.

Figure 6 is an illustrative example of the evolution of final number of cells throughout the treatment schedule given the four re-growth models for 60 days period and $t_2 = 60$ days.

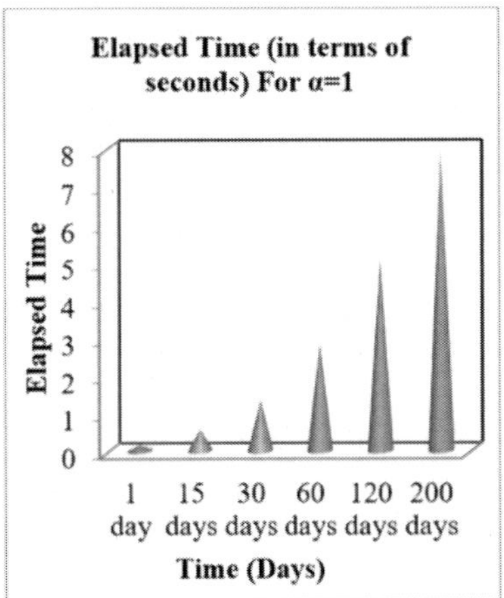

(a)

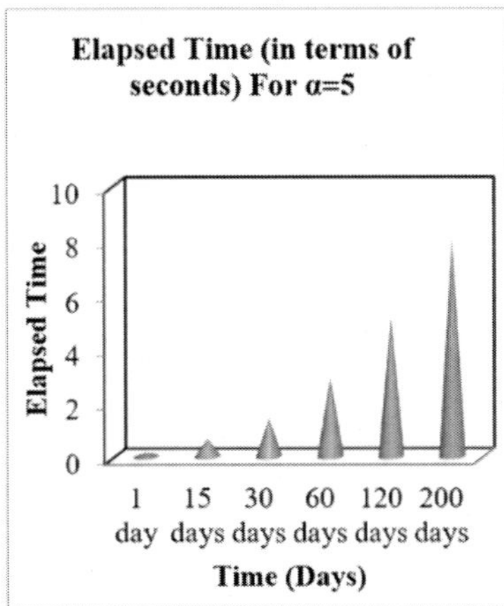

(b)

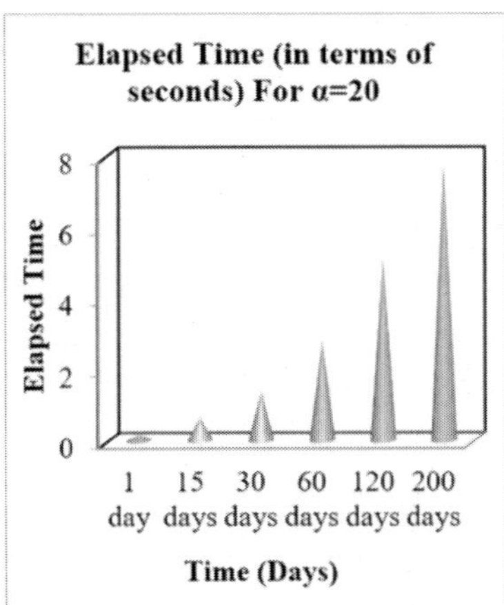

(c)

Figure 7. Total Elapsed time (in seconds) for all computations when (a) $\alpha = 1$ Gy^{-1} (b) $\alpha = 5$ Gy^{-1} (c) $\alpha = 20$ Gy^{-1}.

To Figure 6, exponential and logistic models give similar results in tumor eradication, restricted exponential and Gompertz models indicate poorer prognosis for tumor eradication than exponential and logistic models.

The computational speed of the macro is investigated by calculating the total elapsed time (in seconds) for various time periods (in days). Figure 7 demonstrates it for $\alpha = 1$ Gy^{-1}, $\alpha = 5$ Gy^{-1} and $\alpha = 20$ Gy^{-1}. The various values of α do not change the elapsed time. An increase in time period leads to an increase in elapsed time. Nevertheless, the maximum elapsed time corresponding to 200 days period is around only 8 seconds for different α values.

DISCUSSIONS AND CONCLUSION

A calculator named as TGTSCalculator for oncological computations has been designed in VBA for the first time in the literature. Computations were performed for all types of tumor growth models and survival fractions for treatment schedules in radiotherapy simultaneously.

This calculator is so practical that users only click CALCULATE button and enter information required in UserForm1 to obtain the results of all computations with graphical displays efficiently without entering any data into the spreadsheet.

TGTSCalculator can be employed as a practical tool for users dealing with oncology and also educational purposes in biology.

TGTSCalculator also gives users an opportunity of not spending most of the time with computations and memorization of complicated formulations.

Lastly, as a future work, this calculator can be adapted to any modifications to be used in computations for chemotherapy and other treatments.

REFERENCES

Achuthan, R., K. Grover, and J. MacFie. 2006. 'A Critical Evaluation of the Electronic Surgical Logbook.' *BMC Medical Education* 6 (1): 15.

Anbazhagan, Ramswamy. 2003. 'Microarray Data Assembler.' *Bioinformatics* 19 (1): 157–158.

Aoki, Y., M. Hashiyada, A. Morioka, M. Nata, and K. Sagisaka. 1997. 'Spreadsheets of a Conventional Application Software for Calculation of Plausibility of Paternity: Application to Parentage Testing with Highly Polymorphic Markers in Deceased Party.' *Nihon Hoigaku Zasshi = The Japanese Journal of Legal Medicine* 51 (3): 196–204.

Brocks, Dion R. 2015. 'USIMPK. An Excel for Windows-Based Simulation Program for Instruction of Basic Pharmacokinetics Principles to Pharmacy Students.' *Computer Methods and Programs in Biomedicine* 120 (3): 154–163.

Brown, Angus M. 1999. 'A Methodology for Simulating Biological Systems Using Microsoft Excel.' *Computer Methods and Programs in Biomedicine* 58 (2): 181–190.

———. 2000. 'Simulation of Axonal Excitability Using a Spreadsheet Template Created in Microsoft Excel.' *Computer Methods and Programs in Biomedicine* 63 (1): 47–54.

———. 2001. 'A Step-by-Step Guide to Non-Linear Regression Analysis of Experimental Data Using a Microsoft Excel Spreadsheet.' *Computer Methods and Programs in Biomedicine* 65 (3): 191–200.

———. 2006. 'A Non-Linear Regression Analysis Program for Describing Electrophysiological Data with Multiple Functions Using Microsoft Excel.' *Computer Methods and Programs in Biomedicine* 82 (1): 51–57.

———. 2010. 'A Spreadsheet Template Compatible with Microsoft Excel and IWork Numbers That Returns the Simultaneous Confidence Intervals for All Pairwise Differences between Multiple Sample Means.' *Computer Methods and Programs in Biomedicine* 98 (1): 76–82.

Carlson, David J., Robert D. Stewart, X. Allen Li, Kristofer Jennings, Jian Z. Wang, and M. Guerrero. 2004. 'Comparison of in Vitro and in Vivo α/β Ratios for Prostate Cancer.' *Physics in Medicine & Biology* 49 (19): 4477.

Chudej, Kurt, Lisa Wagner, and Hans Josef Pesch. 2015. 'Numerical Solution of an Optimal Control Problem in Cancer Treatment: Combined Radio and Anti-Angiogenic Therapy.' *IFAC-PapersOnLine* 48 (1): 665–666.

Conway, Tyrrell, Barbara Kraus, Don L. Tucker, Darren J. Smalley, Angela F. Dorman, and Loretta McKibben. 2002. 'DNA Array Analysis in a Microsoft® Windows® Environment.' *Biotechniques* 32 (1): 110–119.

Craighead, Peter S., Khalid Sait, Gavin C. Stuart, Keith Arthur, Jill Nation, Maire Duggan, and Dianlin Guo. 2000. 'Management of Aggressive Histologic Variants of Endometrial Carcinoma at the Tom Baker Cancer Centre between 1984 and 1994.' *Gynecologic Oncology* 77 (2): 248–253.

Dansirikul, Chantaratsamon, Malcolm Choi, and Stephen B. Duffull. 2005. 'Estimation of Pharmacokinetic Parameters from Non-Compartmental Variables Using Microsoft Excel®.' *Computers in Biology and Medicine* 35 (5): 389–403.

Dixit, Deep Shikha, Deepak Kumar, Sanjeev Kumar, and Rajesh Johri. 2012. 'A Mathematical Model of Chemotherapy for Tumor Treatment.' *Advances in Applied Mathematical Biosciences* 3: 1–10.

Enderling, Heiko, and Mark AJ Chaplain. 2014. 'Mathematical Modeling of Tumor Growth and Treatment.' *Current Pharmaceutical Design* 20 (30): 4934–4940.

Geeraerd, A. H., V. P. Valdramidis, and J. F. Van Impe. 2005. 'GInaFiT, a Freeware Tool to Assess Non-Log-Linear Microbial Survivor Curves.' *International Journal of Food Microbiology* 102 (1): 95–105.

Gilani, A. Ghanadzadeh, M. Moghadam, and M. S. Zakerhamidi. 2011. 'Dimeric Spectra Analysis in Microsoft Excel: A Comparative Study.' *Computer Methods and Programs in Biomedicine* 104 (2): 175–181.

Glazebrook, R., B. Chater, P. Graham, G. Kokar, M. Robinson, R. Escott, J. Putman, and D. Crosbie. 2005. 'Evaluation of the ACRRM National Radiology Program for Australian Rural and Remote Medical Practitioners.' *Rural & Remote Health* 5 (3): 349–356.

Gomes, Petrus Pereira, Luis Augusto Passeri, and José Ricardo de Albergaria Barbosa. 2006. 'A 5-Year Retrospective Study of Zygomatico-Orbital Complex and Zygomatic Arch Fractures in Sao Paulo State, Brazil.' *Journal of Oral and Maxillofacial Surgery* 64 (1): 63–67.

Greensmith, David J. 2014. 'Ca Analysis: An Excel Based Program for the Analysis of Intracellular Calcium Transients Including Multiple, Simultaneous Regression Analysis.' *Computer Methods and Programs in Biomedicine* 113 (1): 241–250.

Hahnfeldt, Philip, Dipak Panigrahy, Judah Folkman, and Lynn Hlatky. 1999. 'Tumor Development under Angiogenic Signaling: A Dynamical Theory of Tumor Growth, Treatment Response, and Postvascular Dormancy.' *Cancer Research* 59 (19): 4770–4775.

Joiner, Michael C., and Albert Van der Kogel. 2016. *Basic Clinical Radiobiology*. Vol. 1. CRC press.

Khan, Haseeb Ahmad. 2004. 'ArraySolver: An Algorithm for Colour-Coded Graphical Display and Wilcoxon Signed-Rank Statistics for Comparing Microarray Gene Expression Data.' *Comparative and Functional Genomics* 5 (1): 39–47.

———. 2005. 'ArrayVigil: A Methodology for Statistical Comparison of Gene Signatures Using Segregated-One-Tailed (SOT) Wilcoxon's Signed-Rank Test.' *Journal of Molecular Biology* 345 (4): 645–649.

———. 2006. 'SCEW: A Microsoft Excel Add-in for Easy Creation of Survival Curves.' *Computer Methods and Programs in Biomedicine* 83 (1): 12–17.

King, Christopher R., Thomas A. DiPetrillo, and David E. Wazer. 2000. 'Optimal Radiotherapy for Prostate Cancer: Predictions for Conventional External Beam, Imrt, and Brachytherapy from Radiobiologic Models.' *International Journal of Radiation Oncology* Biology* Physics* 46 (1): 165–172.

Kirschner, Denise, and John Carl Panetta. 1998. 'Modeling Immunotherapy of the Tumor–Immune Interaction.' *Journal of Mathematical Biology* 37 (3): 235–252.

Lambert, Ronald JW, Ioannis Mytilinaios, Luke Maitland, and Angus M. Brown. 2012. 'Monte Carlo Simulation of Parameter Confidence Intervals for Non-Linear Regression Analysis of Biological Data Using Microsoft Excel.' *Computer Methods and Programs in Biomedicine* 107 (2): 155–163.

McAneney, H., and S. F. C. O'Rourke. 2007. 'Investigation of Various Growth Mechanisms of Solid Tumour Growth within the Linear-Quadratic Model for Radiotherapy.' *Physics in Medicine & Biology* 52 (4): 1039.

Meineke, Ingolf. 2000. 'An Add-in Implementation of the RESAMPLING Syntax under Microsoft EXCEL.' *Computer Methods and Programs in Biomedicine* 63 (2): 99–104.

Messmer, Bradley T. 2005. 'SAIVGeM: Spreadsheet Analysis of Immunoglobulin VH Gene Mutations.' *BioTechniques* 39 (3): 353–358.

Panetta, John Carl, and J. Adam. 1995. 'A Mathematical Model of Cycle-Specific Chemotherapy.' *Mathematical and Computer Modelling* 22 (2): 67–82.

Petrakis, Leonidas A., Dorothea A. Petraki, and Andreas L. Petrakis. 2017. 'A Predictor-Corrector Method with a Non Periodic Treatment Factor for the Mathematical Modelling of Tumour Growth.' *Journal of Interdisciplinary Mathematics* 20 (8): 1679–1690.

Pfaffl, Michael W., Ales Tichopad, Christian Prgomet, and Tanja P. Neuvians. 2004. 'Determination of Stable Housekeeping Genes, Differentially Regulated Target Genes and Sample Integrity: BestKeeper–Excel-Based Tool Using Pair-Wise Correlations.' *Biotechnology Letters* 26 (6): 509–515.

Rathod, P. S., P. V. Reddihalli, S. Krishnappa, U. K. Devi, and U. D. Bafna. 2014. 'A Retrospective Clinicopathological Study of 131 Cases with Endometrial Cancers-Is It Possible to Define the Role of

Retroperitoneal Lymphadenectomy in Low-Resource Settings?' *Indian Journal of Cancer* 51 (1): 54.

Rejniak, Katarzyna Anna, and Heiko Enderling. 2013. *Computational Models in Oncology: From Tumor Initiation to Progression to Treatment*. Switzerland: Frontiers Media SA.

Rocheta, Margarida, F. Miguel Dionísio, Luís Fonseca, and Ana M. Pires. 2007. 'Paternity Analysis in Excel.' *Computer Methods and Programs in Biomedicine* 88 (3): 234–238.

Sachs, R. K., L. R. Hlatky, and P. Hahnfeldt. 2001. 'Simple ODE Models of Tumor Growth and Anti-Angiogenic or Radiation Treatment.' *Mathematical and Computer Modelling* 33 (12–13): 1297–1305.

Sato, Hitoshi, Satomi Sato, Ya Min Wang, and Isamu Horikoshi. 1996. 'Add-in Macros for Rapid and Versatile Calculation of Non-Compartmental Pharmacokinetic Parameters on Microsoft Excel Spreadsheets.' *Computer Methods and Programs in Biomedicine* 50 (1): 43–52.

Schageman, J. J., M. Basit, T. D. Gallardo, H. R. Garner, and R. V. Shohet. 2002. 'MarC-V: A Spreadsheet-Based Tool for Analysis, Normalization, and Visualization of Single CDNA Microarray Experiments.' *Biotechniques* 32 (2): 338–344.

Scribner Jr, Dennis R., Joan L. Walker, Gary A. Johnson, Scott D. McMeekin, Michael A. Gold, and Robert S. Mannel. 2001. 'Laparoscopic Pelvic and Paraaortic Lymph Node Dissection: Analysis of the First 100 Cases.' *Gynecologic Oncology* 82 (3): 498–503.

Thames, H. D., S. M. Bentzen, I. Turesson, M. Overgaard, and Walter Van den Bogaert. 1990. 'Time-Dose Factors in Radiotherapy: A Review of the Human Data.' *Radiotherapy and Oncology* 19 (3): 219–235.

Wera, Stefaan. 1998. 'An EXCEL-Based Method to Search for Potential Ser/Thr-Phosphorylation Sites in Proteins.' *Computer Methods and Programs in Biomedicine* 58 (1): 65–68.

Wheldon, T. E., and J. Kirk. 1976. 'Mathematical Derivation of Optimal Treatment Schedules for the Radiotherapy of Human Tumours. Fractionated Irradiation of Exponentially Growing Tumours.' *The British Journal of Radiology* 49 (581): 441–449.

Wheldon, T. E., J. Kirk, and J. S. Orr. 1977. 'Optimal Radiotherapy of Tumour Cells Following Exponential-Quadratic Survival Curves and Exponential Repopulation Kinetics.' *The British Journal of Radiology* 50 (597): 681–682.

Wu, S. 2004. 'Simulations of the Cardiac Action Potential Based on the Hodgkin-Huxley Kinetics with the Use of Microsoft Excel Spreadsheets.' *Chinese Journal of Physiology* 47 (1): 15.

Yankeelov, Thomas E., Nkiruka C. Atuegwu, Natasha G. Deane, and John C. Gore. 2010. 'Modeling Tumor Growth and Treatment Response Based on Quantitative Imaging Data.' *Integrative Biology* 2 (7–8): 338–345.

Zhang, Yong, Meirong Huo, Jianping Zhou, and Shaofei Xie. 2010. 'PKSolver: An Add-in Program for Pharmacokinetic and Pharmacodynamic Data Analysis in Microsoft Excel.' *Computer Methods and Programs in Biomedicine* 99 (3): 306–314.

Zou, Jiaqi, and Na Li. 2013. 'Quantitative Thermodynamic Predication of Interactions between Nucleic Acid and Non-Nucleic Acid Species Using Microsoft Excel.' *Computer Methods and Programs in Biomedicine* 111 (3): 755–762.

In: Human-Computer Interaction
Editor: Turgut Ozseven
ISBN: 978-1-53616-495-4

Chapter 4

THE USE OF INTERPOLATION METHODS FOR MODELLING HUMAN BODY MOTION

Egemen Halıcı[1], Erkan Bostancı[1], Sevgi Yiğit-Sert[1], Mehmet Serdar Güzel[1,*] and Nadia Kanwal[2]

[1]Department of Computer Engineering, Ankara University, Ankara, Turkey

[2]Lahore College for Women University, Lahore, Pakistan

ABSTRACT

Traditional human motion modelling algorithms employ all body joint information to model human motion. In this study, we propose to use some parts instead of all data for modelling human motion. We utilize interpolation methods such as Lagrange, Spline and Cubic Spline interpolations to predict unknown data points. The data points of joint are obtained by Microsoft Kinect Sensor and OpenPose algorithm. For each body joint position, a polynomial is obtained, and then unknown data points are predicted. Experiments show that the proposed method is more memory-efficient and requires low resource and also yields satisfactory results.

[*] Corresponding Author's Email: mguzel@ankara.edu.tr.

Keywords: interpolation, human motion modelling, dynamic movement primitives, Microsoft Kinect, OpenPose

INTRODUCTION

Robotic systems are frequently used in actions where human power is insufficient, or that are harmful to human health, or in operations that require time saving. Besides, non-stop systems and customized systems developed for specific jobs provide substantial advantages in the industry. Autonomous systems have been used in textile industry since the 18th century and their usage has also spread to other sectors over time.

The word "robot" was first used in a play called "Universal Robots" written by Czechoslovak writer Karel Čapek. Generally, it defines autonomous or pre-programmed electronic devices.

Nowadays, with the development of technology, scientists have started to build humanoid robots (Inoue et al. 2000). The process of embedding human mobility capabilities to robots in advance was high-costly, time-consuming and error-prone. However, Dynamic Movement Primitives (DMP) enables us to model and transfer human movements to humanoid robots.

Human movements occur as a result of changes in body joint positions. The speed, position and acceleration of the joints can be calculated by the changes in limbs during the movement. Data sets are generated by following these changes.

DMP are the mathematical expression of the shapes generated by the human skeleton (Prada et al. 2013). The modelling of the skeletal system could be done "monitoring the human movements with DMP" (Ben Amor et al. 2014). The transfer of these models to robots provides that humanoid robots can act human movements without any preparation in advance (Inoue et al. 2000). Since the all of data points is used during digitization of human movements as data set, DMP requires high computing power.

There are numerous ways to model human movements in the literature. Policy Learning by Weight Exploration with Return (POWER), Supporting

Vector Machine (SVM), unsupervised machine learning algorithm are widely used for modelling (Kober and Peters 2009, Arai and Murakami 2018, Sidenbladh 2004, Srivastava 2015, Yang and Goncalves 2003). These techniques can also be used to classify movements. In these studies, it is also seen that the modelling process employs the entire data set (Joo, Simon, and Sheikh 2018).

OpenPose and Microsoft Kinect construct data sets by predict human skeleton at a certain distance (Cao et al. 2018; Kar 2010). In traditional human skeletal tracking systems, data sets are formed with at least two different cameras by attaching reflective markers on the joints. Microsoft Kinect device and OpenPose algorithm are preferred to minimize the cost and system requirements. The differences between the tracking accuracies of Microsoft Kinect and OpenPose algorithms were compared in a previous study (Arduengo Garcia et al. 2017).

In this study, we aim to model human movements using specific data points instead of using entire dataset as in DMP algorithms by employing interpolation methods. Interpolation methods enable us to detect more effective and appropriate data points instead of using all data points for modelling. This provides us to model human movements more accurately with lower computer effort.

The data set is created by tracking the human movements using the OpenPose algorithm and the Microsoft Kinect device. Three different interpolation methods are used to model the data set. The most effective interpolation method is determined by comparing the predicted data from models with the actual data (Halıcı and Bostanci 2016 at ICAA'16).

Although OpenPose algorithm and the Kinect device operate in a similar way, they can create different data sets even for the same movement to be modelled. The reason for these differences is usually that the joints overlap. We project that these missing data points can be completed with the proposed method.

The dance exists from quite a long human history and is an important part of human life. Thus, in this study, we used the data set that is obtained from dance figures. We modelled dance to be transferred to robots, so that robots can acting as human movements. In addition, this study might be used in the dance training. For example; the instructor and the student's dance movements at different times can be compared to assess student's performance.

Lagrange, Spline and Cubic Spline interpolations are used while formulating of polynomials (Halıcı and Bostanci 2016 at SIU16). The error is calculated by using the Mean Squared Error (MSE) and Mean Absolute Error (MAE). During experiments, we focus on generating more accurate models by reducing the error rates. The results are analyzed by ANOVA and McNemar statistical test. Finally, the performances of models are discussed.

MODELLING USER MOTION DATA

In this section; the approach followed for modelling the motion data obtained using two different acquisition methods is described. These approaches can be briefly summarized as follows:

- The approach used to acquire data using two methods, namely: Microsoft Kinect and OpenPose algorithms,
- Modelling the data using optimal step and group values in order to minimize the computational expense and minimize prediction error (mean absolute error and mean squared error) for data points which are not included in the model,
- Finally, a statistical test was performed in order to identify the statistical significance of the performances of different mathematical model in the prediction of human motion data.

Obtaining Data Set with Microsoft Kinect Sensor and OpenPose Application

Kinect is a Microsoft product developed to follow the human skeletal system. It provides an interactive framework for players of the XBox game console. Additionally, it enables software developers to code their own games with Windows Software Development Kit (SDK) and data transfer cable (Zerpa et al. 2015). The sensor on Kinect processes a maximum of 30 frames per second for tracking human movements and obtains data from 20 different body joint on the skeletal system. It is possible to make less frame tracking per second. The value of Frame per Second (FPS) can be reduced to fewer than 30.

The position values of the joints on the human skeletal system are obtained from each frame by processing images taken with Kinect. The obtained joint position values are recorded in files to be used in modelling.

Unlike OpenPose, the joint positions obtained with Kinect sensors are in amc format and the numbers of frames are constant.

Table 1 shows the data set of joint position values on the Frame 270 which is any frame throughout human movements tracked by Kinect sensors. The data set is separated by spaces, and begins with the frame number which it belongs and followed by joint name, X position, Y position and rotation information.

In the case that the movements are very fast, data losses occur while tracking of Kinect sensors. Human movements are monitored by the depth camera on the device. The camera computes position values by making predictions on body joints. Nevertheless, data loss might yet occur due to overlapping of the joint points during movement. The proposed algorithm solves this problem.

A Microsoft Kinect Sensor (V1) tracks up to 2 different skeletons at the same time. The number of skeletons (i.e., 1 or 2) has no effect over the performance of the system. Multiple skeletal tracking results in loss of data because of overlapping body joints of different skeletons according to camera angle (Kar 2010, Alexiadis et al. 2011, Shotton et al. 2011).

OpenPose is open source software that models human movements tracked on video, image and instant recording. It monitors instantly human body joints, face and hand movements frame by frame. During tracking, the distance between the camera and human to be modelled needs to be constant. Especially when modelling through videos, deviations arise in the records that are zoomed in since overall skeleton is not tracked.

Table 1. The data of joint position values obtained by Microsoft Kinect Sensor

Frame #270			
HipCenter	1.517299	-0.2663365	3.172178
Spine	1.527393	-0.2054516	3.178674
ShoulderCenter	1.577862	0.09897247	3.225849
Head	1.502417	0.2761515	3.186341
ShoulderLeft	1.478438	0.02340087	3.370244
ElbowLeft	1.429889	-0.1756384	3.362745
WristLeft	1.290652	-0.1238605	3.208721
HandLeft	1.257245	-0.1065912	3.16587
ShoulderRight	1.633906	0.02130159	3.099792
ElbowRight	1.625241	-0.2493066	3.012941
WristRight	1.443323	-0.1537324	2.998887
HandRight	1.386791	-0.1517075	2.991777
HipLeft	1.471184	-0.3330276	3.21974
KneeLeft	1.470582	-0.7633861	3.124828
AnkleLeft	1.465996	-1.067358	3.100488
FootLeft	1.430762	-1.151783	3.070029
HipRight	1.541144	-0.3339723	3.10046
KneeRight	1.54665	-0.6347807	2.968357
AnkleRight	1.579809	-0.8553371	2.917449
FootRight	1.536089	-0.9351836	2.949226

As mentioned before, multiple people can be monitored during modelling. However, depending on the power of the computer, the number of frames processed in a second varies according to the number of people in the video. This case does not pose a problem, since monitoring is performed over the frames in the course of computations.

As can be seen in Table 2 the data set of joints obtained by OpenPose is in JSON format. The first entry in the data set corresponds to the zones marked on Figure 1. In the data set obtained by OpenPose, some joint position values are as (x1, y1, c1, x2, y2, c2), and some joint position values are as (x, y, c). The variable c indicates the confidence score of the data. The existence of more than one x, y, c variable in the joint position information is due to the estimates made for the same joint in the same frame in different confidence scores. The preference of the joint's position information with a higher c value provides a better modelling (Cao et al. 2017).

Table 2. The data set of joints obtained by OpenPose

"0":[310.011, 75.0179, 0.983612, 286.022, 84.0426, 0.757338],	"1":[335.012, 96.0035, 0.74601, 268.035, 109.003, 0.78616],	"2":[327.001, 98.9962, 0.663911, 260.005, 109.007, 0.895391],	"3":[304.985, 134.005, 0.628113, 265.955, 138.968, 0.860017],
"4":[276.048, 107.96, 0.858432, 298.984, 130.007, 0.445422],	"5":[343.003, 92.0392, 0.809188, 278.003, 108.03, 0.59922],	"6":[327.004, 116.044, 0.869507, 275.994, 140.019, 0.0512275],	"7":[289.017, 101.024, 0.992053],
"8":[328.992, 173.021, 0.609511, 269.004, 174.008, 0.647082],	"9":[268.002, 235.976, 0.616006, 335.986, 236.031, 0.70483],	"10":[261.008, 288.001, 0.377169, 341.008, 288.001, 0.373216],	"11":[336.013, 173.02, 0.639251, 280.978, 173.996, 0.540456],
"12":[270.019, 234.997, 0.47316, 324.998, 236.996, 0.721656],	"13":[260.987, 284, 0.322648, 331.004, 291.014, 0.472995],	"14":[283.964, 82.9872, 0.913366],	"15":[313.979, 68.0419, 0.954664, 284.054, 83.9762, 0.132955],
"16":[269.025, 87.0142, 0.756887],	"17":[326.002, 66.9947, 0.892902]		

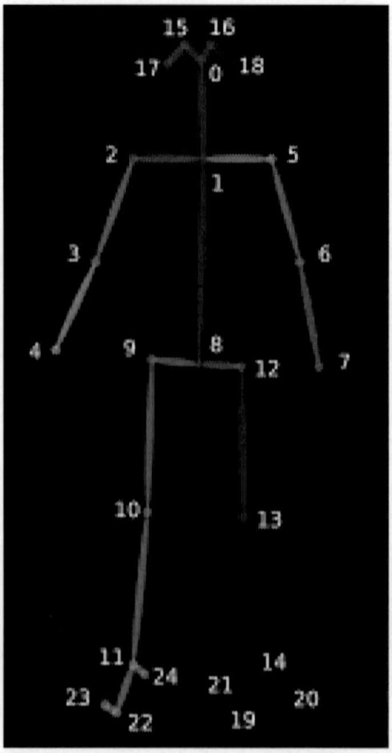

Figure 1. Marked joints on the skeletal system tracked by OpenPose (Arduengo Garcia 2017).

The Dataset Preparation

We obtained the data set of joints by OpenPose in the previous section. Here, we explain how the data is arranged with respect to group and step values. The sampling for Kinect sensors is done from 5-seconds video by extracting 150 frames. When the group value is set as 5 and the step value is set as 8, 18 different frames are used. Four different groups occur since the group value is 5. In this study, each body joint is kept in a separate file. Each group in the same body joint is stored in different files. In the end, the data of a 5-second video are held in 80 different files.

Skeletal tracking was performed from the images taken by OpenPose. Unlike Kinect sensors, there is no fixed FPS value for OpenPose and it varies according to the system features. In this study, we employ OpenPose on the computer with Intel i7 4.00 GHz processor, 32 GB memory, 64-bit

Windows 10 operating system, and NVIDIA GeForce GTX790 4GB 256-Bit graphics card. Since OpenPose runs on the graphics card, the power of the graphics card for the FPS is quite important.

We observed that the average of FPS is around 12-13 while tracking a skeleton with OpenPose that runs on the computer with specified features. Therefore, there are significant differences on the number of frames though the same video is used for Kinect and OpenPose. The sampling for OpenPose generates 60 frames from 5-seconds video. When the group value is set as 5 and the step value is set as 8, 7 different frames are used. Two different groups occur since the group value is 5.

We converted the data that is obtained from Kinect sensors and OpenPose into a common format in order to be compatible. Thus, we facilitate the data to be modelled with the same algorithms even if source of the data is different. In Kinect sensors side, the data starts firstly with the frame number and continues with the body joint names for this frame. The position information for each body joint in the current frame is on the next of joint name. The values of position information are separated by space. A list for each body joint that stores the position information of the joint is created by reading each row of the data. It means that the number of lists is as the number of joints tracked as. The size of each list is equal and is the number of frames obtained during the tracking of the movements. In OpenPose side, data is exported in JSON format. Similarly, a list for each body joint that stores the position information of the joint is created by processing the data.

The Proposed Method

The entire dataset is used in mathematical formulas in DMP algorithm for human motion modelling. The more data set enlarges, the more time consumed for the computation of mathematical operations. In this study, we propose to perform the modelling with interpolation methods to model by using less data so as to reduce time complexity.

Step and group values are required to model data sets for interpolation methods. The increase in the step value reduces the size of the data set used for prediction, while the smaller group value simplifies computations to estimate the joint values from the polynomials. The degree of polynomials in modelling the movement is group value - 1. Thus, the computation time increases incrementally as the group value increases.

Table 3. Data set size differences between DMP and interpolation methods for modelling human movements

Data set size	Step value	Number of Data Points Used for DMP	Number of Data Points Used for Interpolation	Percent of data used in interpolation
100	8	100	12	12%
100	2	100	50	50%
400.000	8	400.000	50.000	12%
400.000	2	400.000	200.000	50%

We choose maximum possible step value and minimum possible group value by examining error rates and so, the heavy workload on computations decreases. In this case, we think that the proposed method will be more efficient modelling than the DMP formulas.

Interpolation

The methods aiming to estimate the value of missing or unknown points from data obtained from the experimental studies are called interpolation (Newman 2006).

Interpolations are usually used for mapping the data points to a function. Thus the unknown or missing points can be estimated from the function. When the $f(x_0), f(x_1), \dots, f(x_n)$ values that a function takes at the finite number of points $x_0, x_1, \dots, x_n \in R$ are known (the function itself is not known), there exists only one n^{th} degree polynomial (see Eq.1).

$$P_n(x) = a_0 + a_1x + a_2x^2 + ... + a_nx^n \tag{1}$$

where ($P_n(x_i) = f(x_i)$ for $i = 0,1,2, ... n.$). When the polynomial $P_n(x)$ is obtained and $P_n(x)$ is replaced with $f(x)$ at a point x, the unknown $f(x)$ value is approximately calculated (Berrut and Trefethen 2004).

In interpolation, polynomials obtained from small fragments obtained by splitting large data sets predict more accurately. For this reason, group and step values are needed at the interpolation to split data set. The group variable defines how many data points are contained in each part, and the step variable specifies which frames in the data set are used. Instead of using the entire data set, the step variable allows modelling with the least amount of data.

The data points that are skipped due to step variable are predicted from polynomials in interpolation process. There are differences between the actual data and data predicted by interpolation methods. These differences are assumed as error of the model and MSE and MAE scores are calculated, so the accuracy of the model is evaluated. The computation of step and group values so that MSE and MAE metrics are minimized leads up to model human motion with the minimum deviation. In this study, Langrange, Linear Spline and Cubic Spline interpolation are used.

Interpolation Calculation with MATLAB

MATLAB performs complex mathematical operations rapidly. There are also built-in mathematical functions that make it easier to implement more complicated operations. On the other hand, functions written by software developers can be converted DLL libraries that enable MATLAB to work compatible with other programming languages.

We converted Langrange, Linear Spline and Cubic Spline interpolation functions in MATLAB to DLL to be used in the rest of study written in C# (Halici and Bostanci 2019). These functions need some parameters in run-time. These parameters should be given from the C# part while using the functions. The MATLAB has its own data types, but these data types can be defined and converted in C# (Moler 2004).

Mean Squared Error (MSE)

The MSE indicates the distance of data points to the regression line. The distances mentioned here are computation errors. The negative signals are eliminated by calculating the squares of the distances, and the mean of the errors is computed, as shown in Eq. 2 (Schluchter, 2014).

$$MSE = \frac{1}{n}\sum_{j=1}^{n}(y_j - \acute{y}_j)^2 \quad (2)$$

Mean Absolute Error (MAE)

The MAE is another metric used to calculate error. Here, the absolute value function is employed in the computation of distances between original and estimated data points to eliminate the negative values, and then the distance values are averaged, as shown in Eq. 3 (Chai and Draxler, 2014).

$$MAE = \frac{1}{n}\sum_{j=1}^{n}\left|y_j - \acute{y}_j\right| \quad (3)$$

Normalization of Datasets

In the comparison of modelled human movements by Microsoft Kinect sensor and OpenPose, the data obtained from people requires to be normalized. In order to compare the models created by the second person as a result of the imitation of the first person, the two datasets must be in the same range of values.

The division of the data of body joint obtained in the normalization by the height of the person who owned the data forms a coefficient. Then, a new dataset is constructed by multiplying the coefficient by the common length which is chosen as 200 for this study. In other words, data points are normalized with respect to a body of 200 cm long.

Let's compute the new value of body joint, x, assuming the body length as b and body joint's position value as e: The equation is shown in Eq. 4.

$$x = \frac{200 * e}{b} \tag{4}$$

Experimental Results

In this study, we modelled the movements of the second person which is the repetition of the dance choreography of the first person. The dance choreography consists of boxing movement figures. Firstly, the models were normalized and the data obtained before and after normalization are shown in Figure 2 and Figure 3.

As can be seen clearly in Figure 2 and Figure 3, the data points exposed to normalization process are moved away from the original data points as expected. The difference is caused by the multiplication of the length 200 cm.

We compared models that are generated on dance movements of two different people. Before the comparison, the starting points of the models were equalized. The aim of this process is to ensure to be the same movements in the same frames.

As two different people perform the movements at different times, different numbers of frames occur in the two data sets. For comparisons, the number of frames is needed to be equal, so the frame number of the data set that has less number of frames are taken into consideration. The missing frames are accepted as error (Halıcı and Bostanci, IJCTE).

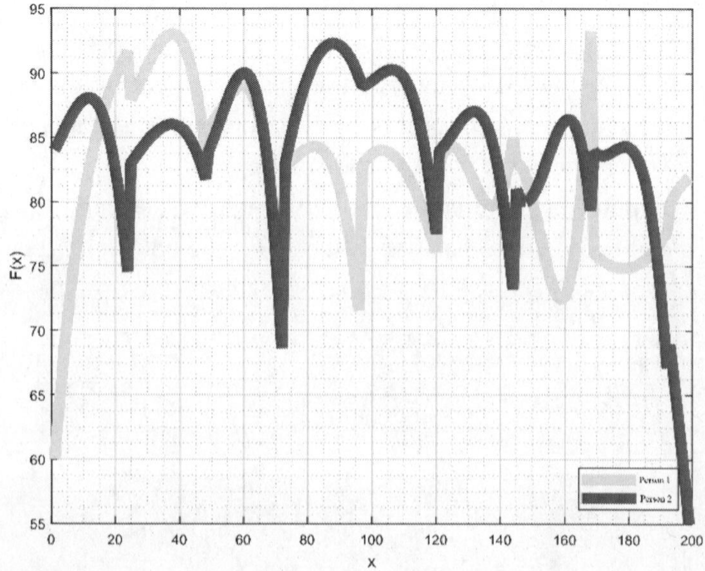

Figure 2. Data points of HipCenter joint values of 1st and 2nd person before normalization. The pink curve displays the 2nd person and the yellow curve displays the 1st person.

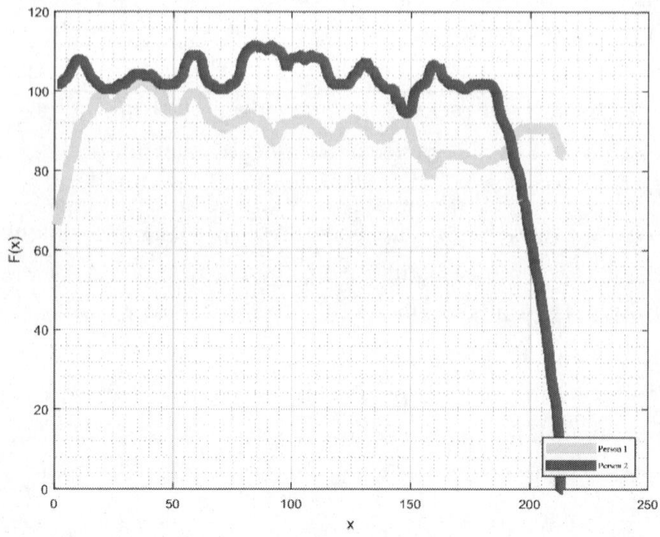

Figure 3. Data points of the HipCenter joint values of the 1st and 2nd person after normalization. The pink curve displays the 2nd person and the yellow curve displays the 1st person.

Table 4. The performance comparison of movements of two different people modelled by Kinect Sensor – OpenPose after normalization

Joint Name	Step and Group Values	Lagrange Success Percentage	Spline Success Percentage	Cubic Spline Success Percentage
Head	6 – 4	10 – 19	11 – 18	8 – 18
Head	2 – 4	17 – 18	13 – 18	15 – 18
ElbowRight	6 – 4	31 – 25	40 – 29	46 – 27
ElbowRight	2 – 4	52 – 26	52 – 27	53 – 27
HipCenter	6 – 4	30 – 10	26 – 17	8 – 17
HipCenter	2 – 4	8 – 16	8 – 17	8 – 17

Table 5. The performance comparison of movements of two different people modelled by Kinect Sensor – OpenPose before normalization

Joint Name	Step and Group Values	Lagrange Success Percentage	Spline Success Percentage	Cubic Spline Success Percentage
Head	6 – 4	31 – 86	25 – 88	29 – 88
Head	2 – 4	33 – 87	38 – 87	33 – 87
ElbowRight	6 – 4	86 – 82	56 – 79	60 – 81
ElbowRight	2 – 4	64 – 81	63 – 82	65 – 82
HipCenter	6 – 4	67 – 88	61 – 89	39 – 90
HipCenter	2 – 4	36 – 89	37 – 89	37 – 37

Table 4 and Table 5 display the performance of different interpolation methods over the motion of two different people modelled by Kinect Sensor and OpenPose before and after normalization, respectively. It is obvious that normalization decreases the performance of all interpolation methods for all body joints. We conclude that body length should not be used in the modelling of human movements. The use of original values of the movements modelled from people is more effective for interpolation performance.

Figure 4 and Figure 5 show the comparison of the information of Head joint that are predicted by Lagrange, Spline, Cubic Spline interpolation methods to the original data which is modelled by Kinect and OpenPose, respectively. The group and step variables for the 1st person are set as 4 and 6. The yellow bold line in the figures denotes the original data. The blue balls, purple crosses and red dashes represent the results of Lagrange, Spline, Cubic Spline interpolation methods, respectively.

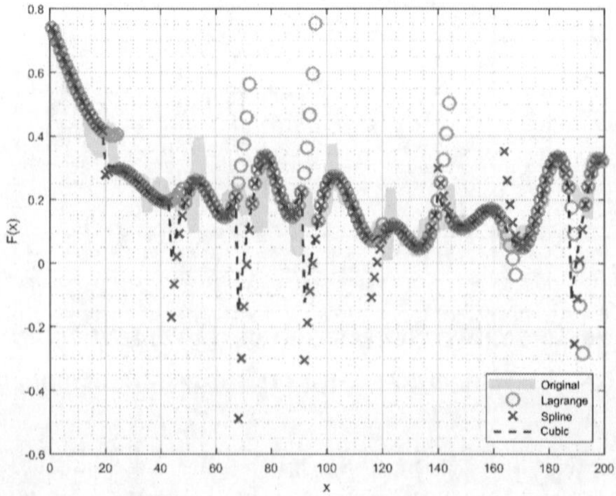

Figure 4. The comparison of the position information of the 2nd person's head joint that is predicted by different interpolation methods when group and step variables for the 1st person modelled by Kinect are set as 4 and 6, respectively.

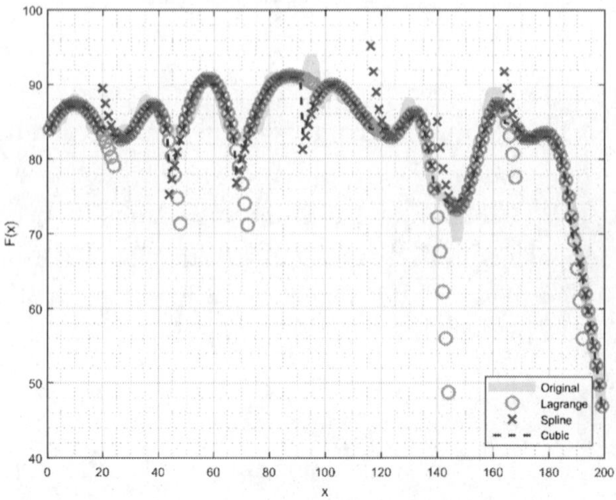

Figure 5. The comparison of the position information of the 2nd person's head joint that is predicted by different interpolation methods when group and step variables for the 1st person modelled by OpenPose are set as 4 and 6, respectively.

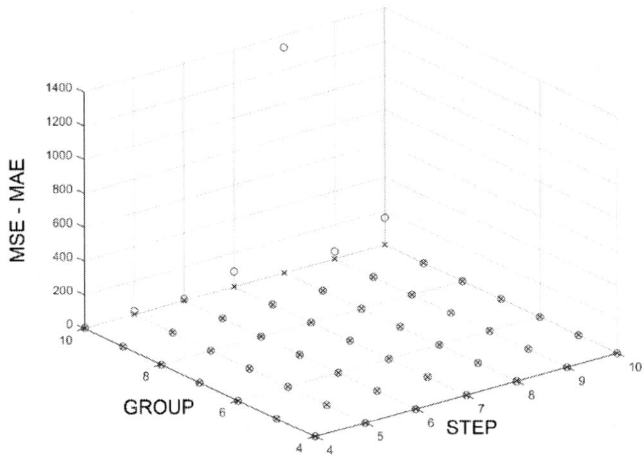

Figure 6. For ElbowRight joint, the MSE and MAE results that are obtained from the comparison of original data of the 1st person modelled by Kinect Sensor and data predicted by Lagrange method.

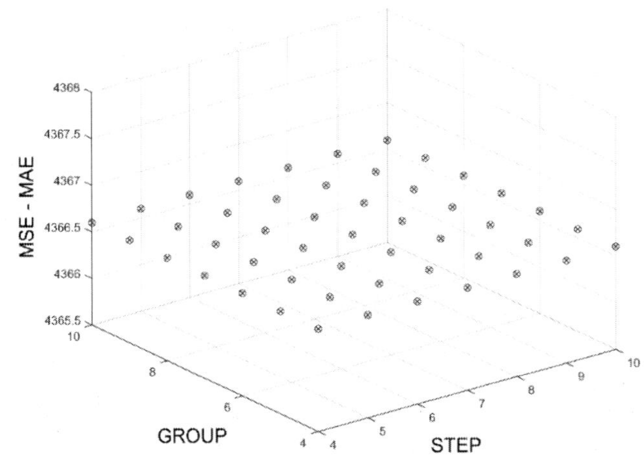

Figure 7. For ElbowRight joint, the MSE and MAE results that are obtained from the comparison of original data of the 1st person modelled by OpenPose and data predicted by Lagrange method.

Figure 6-9 represent the error rates of the Elbow Right joint prediction that is performed by Lagrange interpolation for 1st and 2nd person in the dataset which is obtained by Kinect Sensor and OpenPose. Red balls display error rate in terms of MSE whereas blue crosses display MAE values at different group and step values.

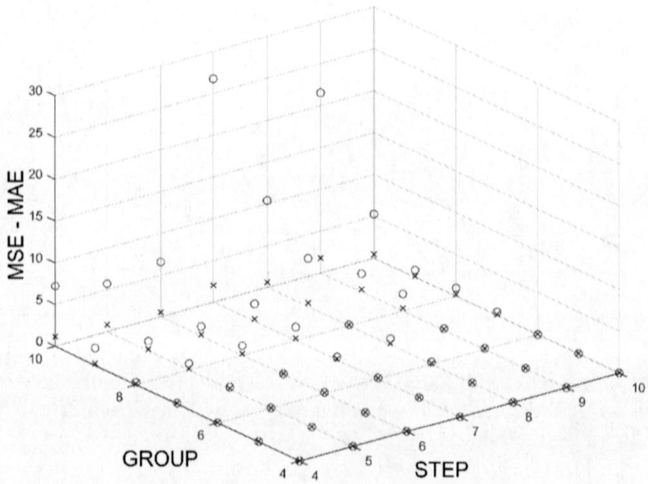

Figure 8. For ElbowRight joint, the MSE and MAE results that are obtained from the comparison of original data of the 2nd person modelled by Kinect Sensor and data predicted by Lagrange method.

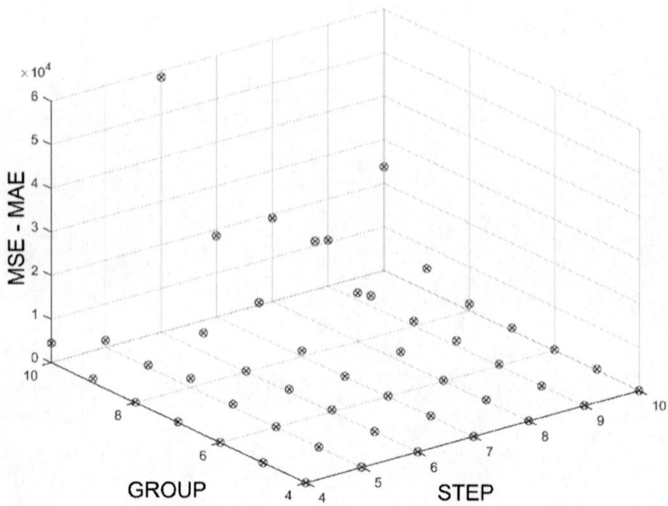

Figure 9. For ElbowRight joint, the MSE and MAE results that are obtained from the comparison of original data of the 2nd person modelled by OpenPose and data predicted by Lagrange method.

Significance Analysis with McNemar Test

In modelling human movements, the positions of the joints on the skeleton are digitized and the digitized data occupies substantial space in storage. In this study, the polynomials are generated from the digitized data, and then the unknown data points are obtained from the generated polynomials. In order to minimize the errors that occur between the predicted data and the original data, the step and group values that fit to modelled movements are selected by calculating the MSE and MAE (Bostanci & Bostanci, 2013).

Table 6 represents variance analysis that is performed by ANOVA test over the generated data set to figure out the performance difference among models that employ different interpolation methods.

Table 6. Variance analysis table for comparison of interpolation methods

Interpolation Method	F	p
Lagrange	15.06	0.002
Spline	4.3	0.001
Cubic Spline	4.45	0

There is a significant difference between interpolation methods as shown in Eq. 5.

$$\begin{aligned} &F_{0,5,\text{ Lagrange Interpolation}} = 15.06,\ p = 0.002 < 0{,}005 \\ &F_{0,5,\text{ Spline Interpolation}} = 4.3,\ p = 0.001 < 0{,}005 \\ &F_{0,5,\text{ Cubic Interpolation}} = 4.45,\ p = 0 < 0{,}005 \end{aligned} \tag{5}$$

Since the interpolation techniques models human movements on the same data, the data is considered as dependent. McNemar test is also convenient for detecting differences in dependent data (Bostanci and Bostanci 2013).

To understand the difference among interpolation techniques, Z score in McNemar test is examined. It is concluded that there is a statistical significance if Z score value is greater than 1.645 in the 95% confidence interval.

Table 7 and Table 8 show McNemar test results on the models of the Head body joint for the data set obtained with Kinect and OpenPose, respectively. The up arrow is drawn to represent that there is a statistical difference or otherwise left-side arrow is drawn.

Table 7. McNemar test results of the error values that occur among the interpolation methods when the range value is 6 and the number of groups is 4 during the modelling of the head joint for the data set obtained with Kinect

Interpolation Method	Lagrange	Spline	Cubic
Lagrange		↑ 1,8195	→ 0.1455
Spline			↑ 14.4836
Cubic			

Table 8. McNemar test results of the error values that occur among the interpolation methods when the range value is 6 and the number of groups is 4 during the modelling of the head joint for the data set obtained with OpenPose

Interpolation Method	Lagrange	Spline	Cubic
Lagrange		→ 0.5801	↑ 9.8884
Spline			↑ 18.8183
Cubic			

CONCLUSION

In this study, we aimed to model human movements using specific data points instead of using entire dataset by employing interpolation methods. Interpolation methods enable us to detect more effective and appropriate data points instead of using all data points for modelling. This provides us to model human motion more accurately with lower computer power. The data set is created by tracking the human motion using the OpenPose algorithm and the Microsoft Kinect sensor. The polynomials are formed for each body joint according to different step and group values. We employed Lagrange, Spline and Cubic Spline interpolation techniques to obtain polynomials from the data. We model human movements and evaluate the performance of models by comparing the predicted data from models with the actual data. Results show that utilizing interpolation in modelling process of human movements yields better performance when compared with other traditional approaches.

REFERENCES

Alexiadis, Dimitrios S., Philip Kelly, Petros Daras, Noel E. O'Connor, Tamy Boubekeur, and Maher Ben Moussa. "Evaluating a Dancer's Performance Using Kinect-Based Skeleton Tracking." In *Proceedings of the 19th ACM International Conference on Multimedia - MM '11,* 2011, 659-662.

Arai, Ryo, Kazuhito Murakami. "Hierarchical human motion recognition by using motion capture system." In *2018 International Workshop on Advanced Image Technology (IWAIT*), 2018, (IEEE, 2018), 1-4.

Arduengo García, Miguel, Steven Jens Jorgensen, Kimberly Hambuchen, Luis Sentis, Francesc Moreno-Noguer, and Guillem Alenyà Ribas. 2017."ROS wrapper for real-time multi-person pose estimation with a single camera. "

Ben Amor, Heni, Gerhard Neumann, Sanket Kamthe, Oliver Kroemer, and Jan Peters. "Interaction Primitives for Human-Robot Cooperation Tasks." In *Proceedings - IEEE International Conference on Robotics and Automation,* 2014, 2831-2837.

Berrut, Jean-Paul, and Lloyd N. Trefethen. 2004. "Barycentric lagrange interpolation." *SIAM Review*, 3: 501-517.

Bostancı Betül, Bostancı Erkan 2013. An evaluation of classification algorithms using Mc Nemar's test BIC-TA 2012.

Cao, Zhe, Tomas Simon, Shih En Wei, and Yaser Sheikh. "Realtime Multi-Person 2D Pose Estimation Using Part Affinity Fields." In P*roceedings - 30th IEEE Conference on Computer Vision and Pattern Recognition*, CVPR, 2017.

Cao, Zhe, Gines Hidalgo, Tomas Simon, Shih-En Wei, and Yaser Sheikh. "OpenPose: realtime multi-person 2D pose estimation using Part Affinity Fields." *arXiv preprint arXiv:1812.08008* (2018).

Chai, T., R. R. Draxler. "Root mean square error (RMSE) or mean absolute error (MAE)?-Arguments against avoiding RMSE in the literature." *Geoscientific Model Development* 7, no. 6 (2014): 1247-1250.

Halıcı Egemen, and Bostanci Erkan. 2019. "Evaluating the Use of Interpolation Methods for Human Body Motion Modelling." *International Journal of Computer Theory and Engineering* 10 ,no. 1 (2019): 25–29.

Halıcı Egemen, and Bostanci Erkan. 2016. Modelling human body motion using lagrange interpolation. *IEEE 24th Signal Processing and Communication Application Conference SIU2016*.

Halıcı, Egemen And Bostancı Erkan. 2016. Analysis Of Training Performances Of Interpolation Methods For Modelling Human Body Motion. 2. *International Conference on Analysis and Its Applications ICAA'16.*

Inoue, Hirochika, Susumu Tachi, Kazuo Tanie, Kazuhito Yokoi, Shigeoki Hirai, Hirohisa Hirukawa, Kazuo Hirai et al. "HRP: Humanoid Robotics Project of MITI."In *International Conference on Humanoid Robots,* 2000.

Joo, Hanbyul, Tomas Simon, and Yaser Sheikh. "Total Capture: A 3D Deformation Model for Tracking Faces, Hands, and Bodies." In *Proceedings of the IEEE Computer Society Conference on Computer Vision and Pattern Recognition*, 2018.

Kar, Abhishek. "Skeletal Tracking Using Microsoft Kinect." *Methodology* 1, no.1 (2010): 1-11.

Kober, Jens, & Peters, Jan R. 2009. "Policy search for motor primitives in robotics." In *Proceedings of the 21st International Conference on*

Neural Information Processing Systems, NIPS, Vancouver, 2008, 849-856

Moler, Cleve B, *Numerical Computing with MATLAB.* (Society for Industrial and Applied Mathematics, 2004), 184.

Newman, Morris. "Interpolation and Approximation." *The American Mathematical Monthly,* 72, no. 8 (1965): 929-30.

Prada, Miguel, Anthony Remazeilles, Ansgar Koene, and Satoshi Endo. "Dynamic Movement Primitives for Human-Robot Interaction: Comparison with Human Behavioral Observation." In *IEEE International Conference on Intelligent Robots and Systems,* Tokyo, 2013, 1168-1175.

Schluchter, M. D. "Mean Square Error." *Wiley StatsRef: Statistics Reference Online*. 2014.

Shotton, J Fitzgibbon, A Cook, M Sharp, T Finocchio, M Moore, R. "Real-Time Human Pose Recognition in Parts from Single Depth Images. In: Computer Vision and Pattern Recognition." In *In Proceedings - 24th IEEE Conference on Computer Vision and Pattern Recognition*, CVPR, 2011.

Sidenbladh, Hedvig. " Detecting human motion with support vector machines ." In *Proceedings of the 17th International Conference on Pattern Recognition*, ICPR, 2004., 188-191.

Song, Yang, Luis Goncalves, and Pietro Perona. " Unsupervised learning of human motion." *IEEE Transactions on Pattern Analysis & Machine Intelligence* 7 (2003): 814-827.

Srivastava, Nitish, Elman Mansimov, and Ruslan Salakhudinov. "Unsupervised learning of video representations using lstms." In *International conference on machine learning*, 2015, 843-852.

In: Human-Computer Interaction
Editor: Turgut Ozseven
ISBN: 978-1-53616-495-4

Chapter 5

A SURVEY OF MOVING TARGET TRAVELING SALESMAN PROBLEM

***Ukbe Uçar*[1,*] *and Selçuk Kürşat İşleyen*[2]**
[1]Department of Industrial Engineering, Fırat University, Elazığ, Turkey
[2]Department of Industrial Engineering,
Gazi University, Ankara, Turkey

ABSTRACT

The Moving-Target Traveling Salesman Problem is an NP-Hard problem that has been extensively studied in recent years. In this problem, it is tried to destroy targets moving at a constant speed by a pursuer within minimum task time. The problem is widely used in many applications such as military operations, scheduling of robotic systems, security camera routing and marine surveillance operations. Furthermore, interest in this problem is expected to grow in parallel with accelerating transition of production systems to Industry 4.0 and the developments in modern warfare technologies. This paper analyzed studies performed on the Moving-Target Traveling Salesman Problem between 1995 and 2019. The examined studies were systematically classified according to their real-life applications, and the solution methodologies used for the problem were also described. Finally, a number of predictions were made

* Corresponding Author's Email: uuucar@firat.edu.tr.

concerning the other versions of the problem and the problem's future areas of application. In addition, this study is known as the first survey study about MTTSP.

Keywords: moving-target traveling salesman problem, aerospace science and technology, dynamic systems, maritime surveillance problem, unmanned aerial vehicles

INTRODUCTION

First studied in the 1930s, the traveling salesman problem (TSP) is an NP-Hard type problem (Sanchit 2010). In the problem the list of cities and the matrix of distance between these cities are given, and the calculation of the shortest tour in which all cities are visited only once by a traveler is aimed. There are several variants of this problem in the literature and there are many studies being conducted within the scope of these problems. Some of the variants of this problems are described below.

Symmetric TSP: A Classical Traveling Salesman Problem in which the distance, or cost, between two nodes is the same in both directions (i.e., moving from node i to j being equal to returning for node j to i) is called a Symmetric TSP. The goal of this problem is to minimize the total tour duration (S.Raja and K. Kannan 2007).

Asymmetric TSP: Distance, or cost, between two nodes is not the same in both directions, then the problem will be called an Asymmetric TSP (In-Chan, Seong-In, and Hak-Soo 2003).

Dynamic TSP: A problem in which parameters such as the duration of travel and the time of arrival change over times is called a Dynamic TSP (Tsung-Sheng, Yat-wah, and Wei Tsang 2009). Pantrigo and Duarte expressed the classic traveling salesman problem as G = {V, E, W} graph, V being the set of nodes showing the cities, E being the set of roads between the cities, and W being the set of weights in the form of symmetric matrix showing the distances between the cities.

And they considered the dynamic TSP as the G graph dependent on time and stated that it is a generic version of the Classic TSP. They stated that the dynamic TSP has two variants, the first being the increase or decrease of the number of the cities and the second being the change of distance between cities in a costant-city network. In both cases, the city location remains constant, only the number of cities or intercity distance change (Juan José and Abraham 2013).

Kinetic TSP: Kinetic TSP: Let S (t) = {s1(t), s2(t), ..., sn(t)} be a set of moving points on an Euclidean metric plane. Let each si (t) point within S(t) set move at the speed of vi <1. Under these assumptions, the kinetic TSP is intended to be the determination of the minimum tour route, which a traveling salesman starts, travels to all points, and returns to the starting point. In this problem, an optimal traveler needs to move at maximum speed, and in the Kinetic TSP the maximum speed is 1. The case where each target in the S set moves at the same speed and in the same direction is expressed as the Translational Traveling Salesman Problem (TTSP) (Mikael and Bengt 2002).

Moving Target Traveling Salesman Problem (MTTSP): Represents a general version of the TSP. The definition and characteristic of this problem are described in section 2 (C.S.Helvig, Gabriel, and Alex 2003).

This study addresses the MTTSP that analyzes the studies conducted between 1995 and 2019 within the scope of the problem in terms of their areas of application. As its areas of application comprise military operations, routing of satellites and ensuring maritime security, this type of problem is of great strategic importance. Other examples of the problem's real-life applications include anti-missile systems, routing of coast guard boasts in seas, refueling of satellites in space, routing and scheduling of robotic systems. In the second chapter of the study, MTTSP is defined and its basic variants are shown. In the third chapter, the studies in the literature are classified according to their application areas and the studies related to the problems associated with MTTSP are touched upon. In the fourth chapter, the solution approaches discussed within the framework of MTTSP are included. In the fifth section, estimates are made about the deficits in the literature and about future studies. In the sixth and final chapter, a general evaluation with respect to the study is conducted.

MOVING TARGET TRAVELING SALESMAN PROBLEM

This section provides a definition for the MTTSP, describing the variants of the problem along with their characteristics. First discussed by Helving et al., the Moving Target Traveling Salesman Problem was described as follows (C.S. Helvig, Gabriel, and Alex 1998).

S = {s1, s2,..., sn} denotes a set of targets in the system. The targets in this set have constant velocity vi, initial position pi. Starting from the origin, the pursuer has a constant velocity of v, with v>|vi|. Based on these assumptions, the Moving Target Traveling Salesman Problem is described as problem of determining the fastest tour route where the pursuer begins at the origin, destroys all targets, and then returns to the origin (C.S. Helvig, Gabriel, and Alex 1998).

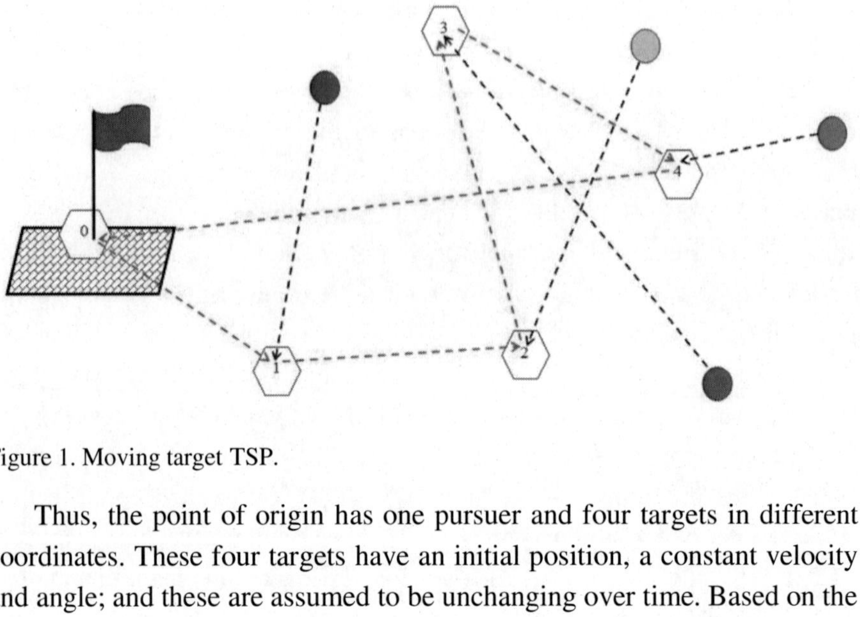

Figure 1. Moving target TSP.

Thus, the point of origin has one pursuer and four targets in different coordinates. These four targets have an initial position, a constant velocity and angle; and these are assumed to be unchanging over time. Based on the illustration in the figure, it can be seen that the pursuer at the point of origin moves to destroy the red target at point one, orange target at point two, blue target at point three and green target at point four, after which it returns to the point of origin. As a result, the pursuer destroys the targets according to the task route of 0-red target-yellow target-blue target-green target-0.

Although it is observed that the MTTSP problem carries to the literature the characteristics similar to the Kinetic TSP and Dynamic TSP, the MTTSP diverges from these problems in various aspects. In MTTSP, the city or node locations move at a certain speed and on a fixed route, and the city locations change over time. And in the dynamic TSP, the locations remain constant, only numbers of the cities and distances between the cities change.

When we compare the MTTSP problem with the Kinetic TSP, we see the following differences. While in the Kinetic TSP speeds belonging to the points are smaller than "1" and maximum speed of the traveler is 1, in the MTTSP the striker and targets may move at different speeds, these speeds are not limited in any way. In addition, while the Kinetic TSP covers the systems that move on a straight line like a conveyor line, the MTTSP tries to catch the targets moving in any direction or line (linear or non-linear).

In the part up to this point, a general description of the MTTSP is made and its relationship with other problems in the literature is explained. And in the following parts, evaluations with respect to the MTTSP's variants, application areas and status in the literature are made.

Variants of the Problem

The MTTSP has a wide range of application area and can be considered from different aspects. The basic variant of the problem is expressed above, and based on this variant, various versions of the problem arise over time. The MTTSP variants identified in the literature are explained in the items below.

Case 1. Problem in Which the Target and Pursuers Move on a One-Dimensional Plane (The X Plane)

In this variant of the problem, the pursuer and targets are assumed to all move in a single direction. Solving the problem involves calculating cost when the pursuer first moves left and then right, as well as the cost in the opposite sequence, in order to select the shortest path(C.S.Helvig, Gabriel, and Alex 2003).

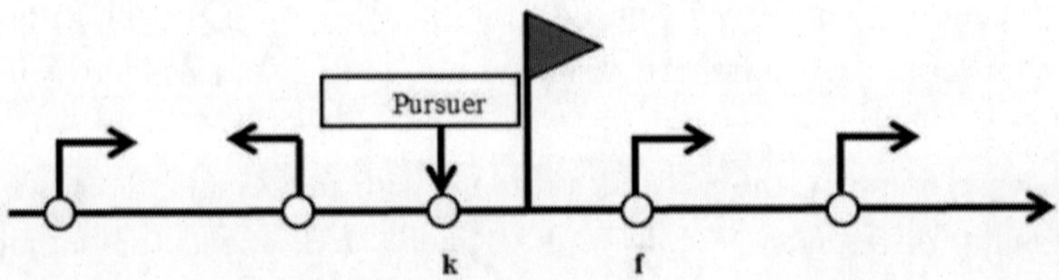

Figure 2. MTTSP in which the target and pursuers move on a one-dimensional plane (the x plane) (C.S.Helvig, Gabriel, and Alex 2003).

Case 2. MTTSP in Which the Pursuer Needs to Return to the Origin for Resupply

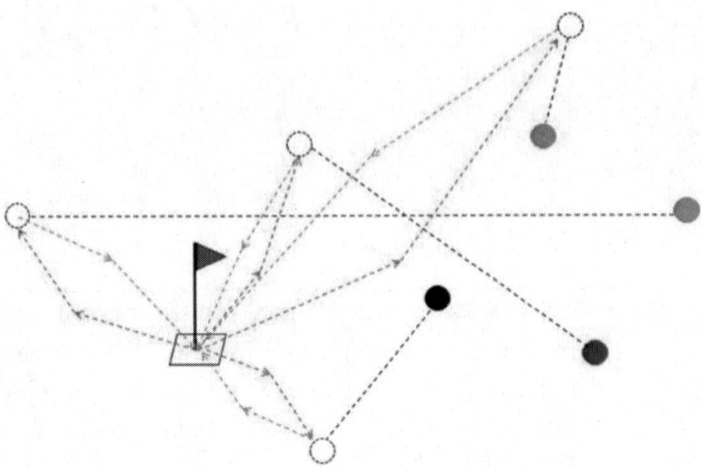

Figure 3. MTTSP in which the pursuer needs to return to the origin for resupply (C.S.Helvig, Gabriel, and Alex 2003).

In this variant, the assumption is that the pursuer needs to return to the origin to resupply after intercepting each target, and that it will move once again to intercept targets after resupplying. The basic assumption of the problem also considers that all targets within the system are moving either towards or away from the origin (C.S.Helvig, Gabriel, and Alex 2003).

Case 3. Problem in Which There Are More Than One Pursuer, and the Pursuers Return to the Origin for Resupply

In this variant of the problem, there are multiple pursuers which need to return to the origin for resupply. The problem is described as having targets moving away from the origin, and the pursuer having an equal velocity of k (C.S.Helvig, Gabriel, and Alex 2003).

Case 4. MMTSP in Which Targets Are Visible Inside the System for Certain Time Window

This variant of the problem, there are multiple pursuers and targets, and targets enter and leave the system within a certain time window. In other words, the number of targets within the system can increase or decrease at any time t. In this problem, all targets need to be intercepted within a certain time period, and the focus is on minimizing the total travel time of the pursuers (Anke et al. 2015; Anke and Armin 2016).

Figure 4 shows two pursuers and 11 targets. The pink and green lines indicate the routes followed by the pursuers. Each black line, on the other hand, indicates a target, while the numbers at beginning and end of these lines indicate the targets' time of entry and exit into and out of the system. The locations where the lines of the pursuers meet with the dots on the black lines show the time points where the target in question was intercepted. For example, looking at Figure 4, it can be understood that the pursuer with pink line intercepted the targets at the time period 14, 16, 17 and 18.

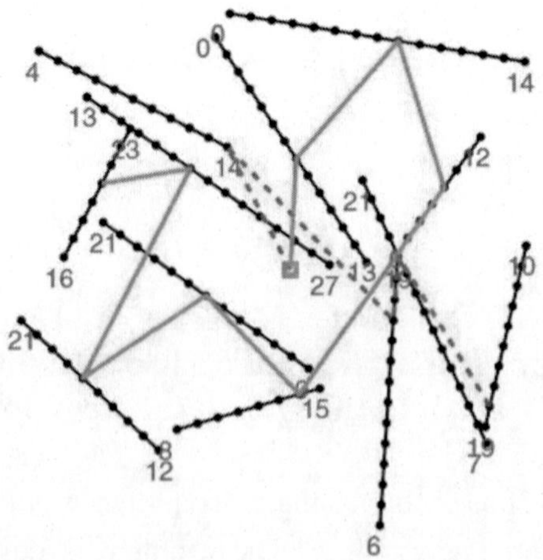

Figure 4. Moving target multiple traveling salesman problem with time window (Anke et al. 2015).

Case 5. Moving Target Multiple Traveling Salesmen Problem with Limited Space

In this variant of the problem, the target and pursuers are confined to a limited mission area, and an effort is made to intercept the targets before they reach a final line (Stephen et al. 2009). The features related to the problem are shown in figure 5 (Stephen et al. 2009).

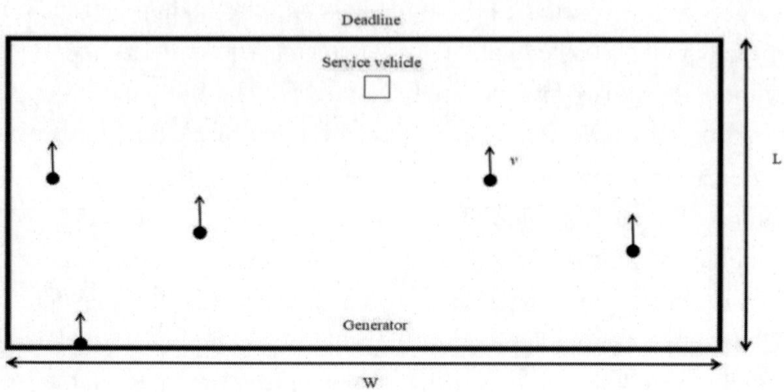

Figure 5. Moving target multiple traveling salesman problem with limited space (Stephen et al. 2009).

In addition to these cases, MTTSP can, based on the definitions in the literature, also be classified according to the cases listed below:

According to the plane in which the pursuer and targets move:

- The target and pursuers move on a one-dimensional plane (x axis).
- The target and pursuers move on a two-dimensional plane (x and y axis).
- The target and pursuers move on a three-dimensional plane (x, y and z axis).

According to the targets' point of arrival:

- Pursuer intercepts all targets and returns to the origin.
- Pursuer intercepts all targets and leaves the system.
- Pursuer intercepts all targets and returns to the origin for resupply.

According to the number of pursuers:

- Single pursuer.
- Multiple pursuers.

According to the velocities of the target and pursuer:

- Pursuer velocities are all equal, while target velocities are different.
- Pursuer velocities are different, while target velocities are all equal.
- Pursuer velocities are different, and target velocities are also different.

According to the level of information on the problem parameters:

- Parameters are deterministic.
- Parameters are stochastic.

It is considered that these classifications for the MTTSP can be also be done differently, and that they can be extended by new parameters.

LITERATURE REVIEW

The literature MTTSP can be examined under different categories according to application areas. The different categories related to the problem are listed as follows in the literature:

- Close Enough Traveling Salesman Problem
- Weapon Target Assignment Problem
- Dynamic Traveling Salesman Problem
- Kinetic Traveling Salesman Problem
- Maritime Surveillance Problem
- Dynamic Vehicle Routing Problem

Based on these categories, studies on the MTTSP were classified under five groups according to their areas of application: These include:

- Military Applications
- Space Applications
- Maritime Activities, Maritime Surveillance Operations, Scanning Activities
- Security Cameras and Surveillance Activities
- Routing and Scheduling of Robotic Systems

The following sections of the article analyses each area of application in detail, and provides a brief summary of the studies in each one of these areas.

Military Applications

One of the most important and strategic areas of application of the Moving Target Traveling Salesman Problem is military applications. Ensuring national security, rapid decision-making at short notice and maximum losses to the enemy while sustaining minimal losses is only possible through the right mission strategy. As the technology advances, the developments in the defense industry are paving the way for unmanned military forces, while also increasing the need for human-machine interaction. Spearheading these advances are missile defense systems, unmanned air vehicles and laser weapons, management and control of these systems representing as an important problem. Some of these systems are shown in Figures 6 and 7. Figure 6 are shown anti drone technology used against drone attacks. Air defense systems that countries use against ballistic missile attacks are indicated in Figure 7.

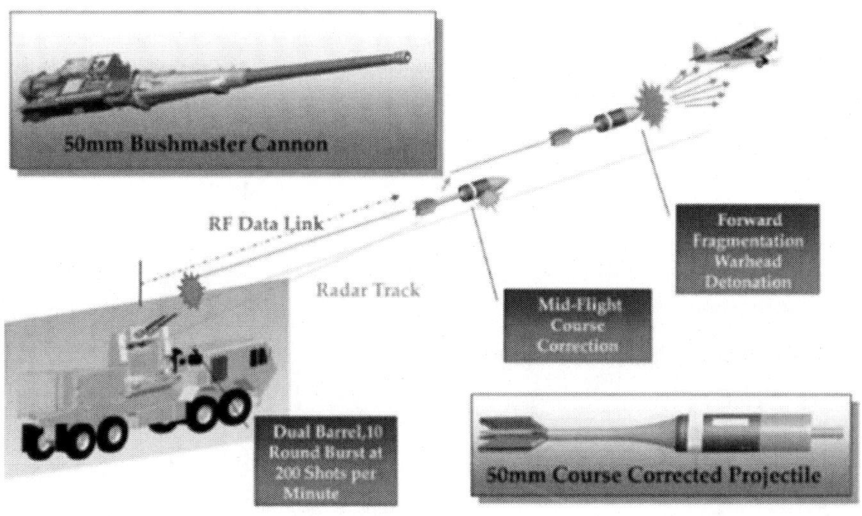

Figure 6. Anti-drone technology (Web1).

The majority of the studies in the literature consist of military applications, and this section of the paper discusses MTTSP studies which have been conducted in military areas.

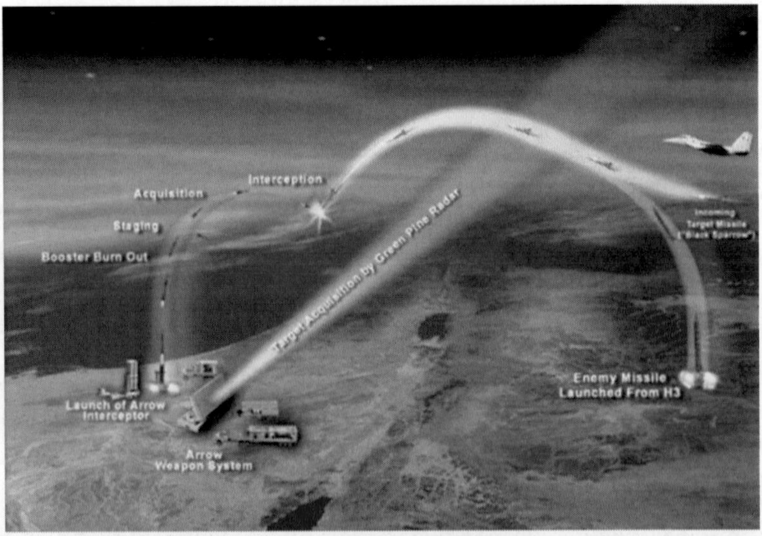

Figure 7. Air defense system used against ballistic missiles (Web2).

The first study on the definition and characteristics of the MTTSP was performed by Helvig et al. This study is also one of the prominent articles concerning this problem's use in military application. In their study, Helvig et al.'s defined a problem for a situation involving a one-dimensional (x) plane and multiple pursuers that resupply, and then described their methods for solving this type of problem (C.S. Helvig, Gabriel, and Alex 1998).

In their study conducted in 2014, Anke et al. developed an integer mathematical model for an MTTSP involving multiple pursuer and targets. The model they proposed also stands out as the first mathematical model to have been developed for the MTTSP. Fügenschuh et al. examined the model based on discrete times, and noted that the model became insufficient when the size of the problem increased. To resolve this issue, they proposed two heuristic algorithms that work based on a minimum time and minimum distance approach. They then tested the proposed solution method on 36 samples in which the moving targets are destroyed by laser weapons, and involving targets and pursuers with differing numbers and velocities. In the objective function, they attempted to minimize the number of tours, and observed that 30 of the trials performed on the 36 samples yielded an optimal solution (Anke et al. 2014).

Stieber et al. considered the multiple traveling salesmen problem with moving targets as a multi-target, multi-pursuer problem, and developed a mathematical model in which time consisted of discrete units. They also stated that in this problem the targets are present in the system within a certain time interval. They also proposed three heuristic algorithms to solve the problem, and tested these algorithms with 36 different sample problems in order to compare them with the mathematical model they developed. In the objective function of the problem, they considered energy consumption as a parameter (Anke et al. 2015).

Stieber and Fugenschuh conducted a study in which the moving target traveling salesman problem was evaluated from a time window perspective. They described that the targets could only be seen within a certain time window, and that their objective was to intercept all targets within this time window, with a minimum number of tours. To be able to provide solutions within an acceptable time period in case the number of variables increases, they proposed to different modeling approaches for this problem. In the first model, they examined the situation when time was considered discrete. In the second model, they developed a model based of Big-M constraints, for the situation in which time was continuous. They noted that in the situation where time was discrete, increasing the number of time steps also considerably increased the time required for obtaining solution, while in the continuous time model, increasing the number of pursuers increased the time need for a solution. Their solution method involved forming sub-problems to resolve this issue. In this context, they first obtained time-independent solutions, and then added time constraints to these solutions, noting that this enabled them to achieve optimal tours (Anke and Armin 2016).

Stieber and Fügenschuh in their study discussed the MTTSP under time window constraint, and unlike other studies in the literature, assumed that the targets move in a nonlinear orbit. They point out that there are a large number of strikers and targets in the problem. In the problem, they considered the time as discrete and developed a mixed integer mathematical model for the solution of the problem. And in conclusion, they stated that the higher the level of interruption in times in the experiments on both linear and nonlinear samples is, the more complex becomes the structure of the problem. (Anke and Armin 2018).

Jiang et al. conducted a study in which they addressed a classical MTTSP involving one pursuer and multiple targets, and in which the velocities and directions of the targets are known. They recommended the genetic algorithm method to solve the problem, using different crossover and mutation operators and the maintenance-repair mechanism to enhance the quality of the solution. They tested this proposed method on 30 different problem variants consisting of 10 cities, while also working to minimize the time of travel in the fitness function. They determined that 18 of the 30 samples yielded definite results, while in the remaining samples, 5 yielded better results with the CX operator, and 7 yielded better results with the OX operator (Q. Jiang, R. Sarker and H. Abbass 2005).

Jindal et al.'s study worked on an MTTSP involving a pursuer that begins its movement in a straight line, and an n number of targets moving at a constant velocity. According to their definition of the problem, the pursuer returned to the origin after intercepting two targets, and they also tried to minimize the time of travel. Furthermore, they developed a heuristic algorithm to solve the problem within the scope of the study. In the algorithm, they positioned the targets on a straight line, and attempted to obtain results by comparing the order of priority assigned to the targets. They determined that the solution involving one pursuer and multiple targets yielded suitable results (Pawan and Amit 2011).

In their study, Englot et al. worked on a MTTSP in which a single pursuer tries to rapidly intercept multiple moving targets. They described that the MTTSP and Kinetic TSP were similar, and developed two different heuristic algorithms to solve the problem. As an applied study, they worked on a problem involving the interception of multiple targets by an unmanned aerial vehicle, and then compared the Greedy Search Methodology with the Lin-Kernighan Heuristic (LKH) in the solution method.

They used agents to instantly monitor the targets, and, with the aid of these agents, they used heuristic algorithms to solve the problem. They described LKH as an agent-based heuristic, and used the agent to determine the pursuer's next point of movement. They also noted that non-greedy heuristics yielded better results in rapidly moving systems, while the LKH yielded better results in slowly moving systems (Brendan, Tuhin, and Isaac 2013).

In their study, Jindal et al. worked on an MTTSP involving the resupply of the pursuer, and proposed two new algorithms to solve the problem. The first algorithm considered the situation where the targets rapidly moved away from the origin at constant velocity, while the second algorithm considered the situation in which all of the targets moved towards the origin on a straight line and at a positive constant velocity. They also assumed that the pursuer resupplied by returning to the origin after intercepting a target, and both algorithms attempted to minimize the total time interception time of the pursuer (Pawal, Amit and Shishir 2011).

Khosravi et al. conducted a study involving a problem with multiple targets to be intercepted, and where these moving targets' routes were indeterminate. The mission area also had certain obstacles and enemy forces. They formulated the problem as an "award-collection mission," and pointed out that they were polygonal obstacles in urban areas. The proposed a heuristic algorithm to solve the problem, and, using a simulation, tested this algorithm on experimental trials (Mohammad and Amir 2014).

Hammar and Nilsson's study evaluated the kinetic version of the TSP, and assumed that each point in the problem moved in a constant direction and constant velocity. They then made deductions concerning the complexity of the problem (Mikael and Bengt 1999).

In their study, Zhou et al. described Dynamic TSP as an NP-Hard type problem, first recommending a mathematical model to solve it, and then proposing the evolutionary algorithm named Dynamic Inver-Over Evolutionary Algorithm (DIOEA) to obtain effective solutions. During optimization, they described their primary goal as finding the tour route that would minimize the time of travel, and then tested the algorithm on a problem consisting of 100 cities, whose locations changed over time according to the Gauss distribution. They compared, both in static and dynamic terms, the changes that occurred over time, and determined the suitable tour routes accordingly. The experimental trials demonstrated that the DIOEA finds best route in real-time (Aimin, Lishan and Zhenyn 2003).

Choubey conducted a study on an MTTSP where targets moved in different directions at constant velocities. In the study, he stated that the targets moved at a certain angle and tried to find the fastest tour where the pursuer would start from the origin and turn back to the original after destroying all the targets. He sequentially worked on problems consisting of 4, 6, 8 and 10 targets, and defined the targets as having a single dimension (a specific distance) and moving at specific velocities. The genetic algorithm developed to solve the problem was applied on various data sets, and the results were compared with those obtained through the greedy heuristic method. In the fitness function of the genetic algorithm, they attempted to minimize the time spent by the pursuer, while in the greedy algorithm, they aimed to determine the cities to be visited according to the distance: velocity ratio. Their study findings demonstrated that the genetic algorithm yielded better results than the greedy algorithm (Nitin 2013).

Lee et al., conducted a study on the weapon-target assignment problem, using the immunity-based ant colony optimization method to solve it. They pointed out that the main objectives in the problem were to assign weapons to targets to minimize anticipated damage, and as a result of the simulation studies, they suggested that the proposed algorithm produces effective solutions (Zne-Jung, Chou-Yuan, and Shun-Feng 2002).

Pushkarini and Bullo performed a study on the dynamic routing problem of moving targets. In this problem, they assumed that the targets are randomly distributed according to the Poisson distribution on a circle with a certain radius, and that they move at constant velocities towards the exterior of the circle. In the objective function, the aimed to maximize the number of targets intercepted within the circle, or unable to escape from the circle. They proposed three different policies to solve the problem, which were the Capturable Nearest Neighbor (CNN), Sector Wise (SW) and Stay Near Boundary (SNB), and observed that the CNN policy yielded better results than the two other policies (Pushkarini and Francesco 2014).

In his study, Mikael and Bengt performed complexity calculations for certain variants of the Kinetic TSP problems, and determined which ones were P type, and which ones were NP type (Mikael and Bengt 2002).

Knapp and Rothe conducted a study on a Weapon-Target Assignment Problem involving the interception by laser weapons of attacks with gun, rocket and mortar munitions. In their study, they analyzed in detail the characteristics, assumptions, threats of the problem, as well as the targeting and interception processes, and proposed a simulation-method based solution to solve the problem (Knapp and Rothe 2012).

Space Applications

Space applications is one of the notable areas where the MTTSP is widely used. The main subject of studies in this area include the control and orientation of satellite systems, and the observation of objects in space. Some of NASA's Earth Science Satellites serving in Earth orbit are shown in Figure 8 and 9 shows an example of how refueling to these satellites is done. This section covers some of the studies from the literature on space applications.

The study of Bourjolly et al. attempted to determine the course of the trip to 20 satellites rotating at a certain angle and steady position around the world by a mobile warehouse in the course of their work, with minimal time and energy use, refueling and returning to the starting point. They calculated the position of the satellites for every minute with a time interval of five years between 2002 and 2007, and assumed the satellites to be moving on a three-dimensional plane. They also developed a comprehensive algorithm to solve small-scale problems, while developing a tabu search algorithm to solve large-scale problems. Based on their results, they described that their proposed method provided suitable tour routes for small- and large-scale problems (Jean-Marie, Ozgur, and Aleksander 2006).

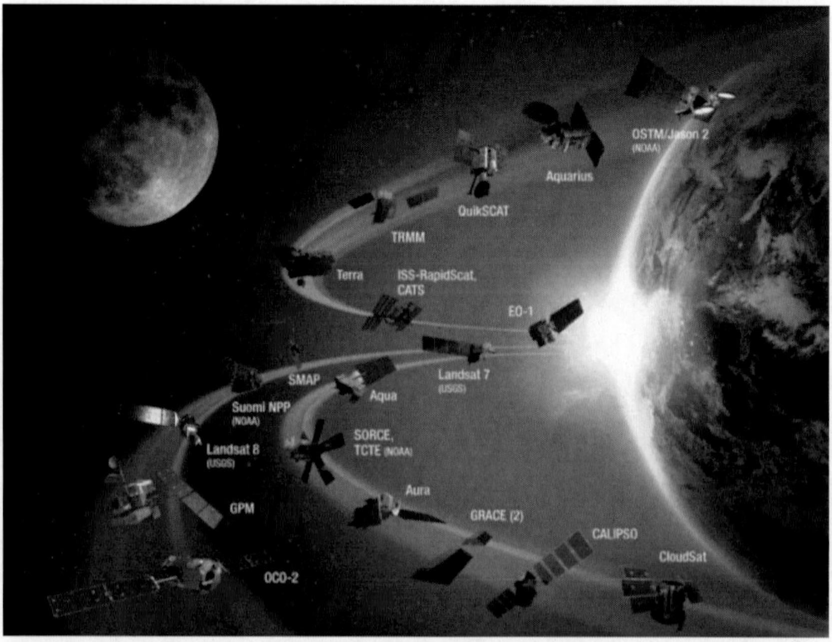

Figure 8. NASA's earth science satellites(Web3).

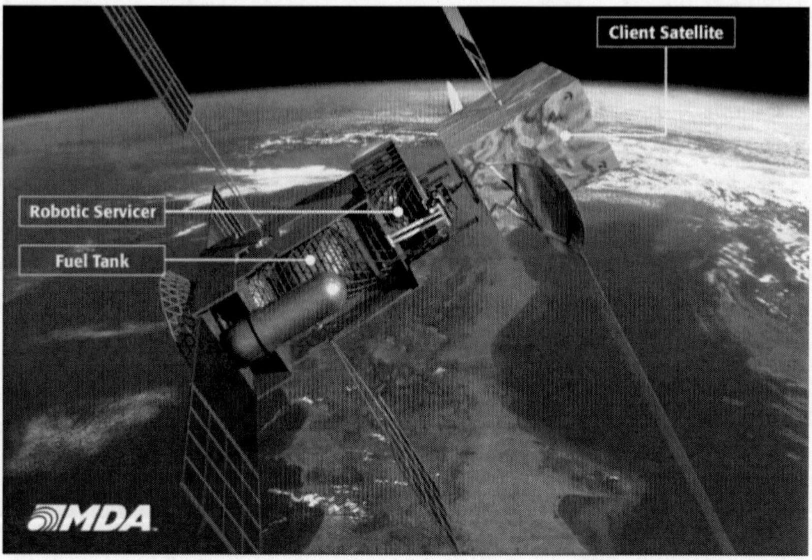

Figure 9. Refueling the Satellites (Web4).

In their study, Blough et al. worked on a problem concerning the routing of a spacecraft that will operate for ten years between 2013 and 2022, by observing and visiting the asteroids located on the L4 and L5 orbits of Jupiter. In this problem, they calculated the distances between the asteroids' orbits, and relied on the Genetic Algorithm and Sequential Meeting Point Method to determine the route resulting in the lowest fuel consumption. Their study results indicated that the Sequential Meeting Point Method yielded better results than the Genetic Algorithm, in terms of the sequence of asteroids to be visited and the duration of flight (Owen, Tyler, and Jennifer 2016).

The study of Mei et al. worked on a problem concerning the observation by satellites of moving targets in space. In this study, to avoid uncertainty resulting from moving targets, it is proposed a scheduling algorithm that measures the information about the targets and evaluates the probability of perception. The validity of the proposed algorithm was tested with the Random Selection Algorithm (RA), the Detection Probability Algorithm (DPA), the Information Algorithm (IA), and the Comprehensive Algorithm (CA). It was found that the proposed algorithm represented an effective method for the detection of the targets, since it took into account both the information on the moving targets and their probability of detection (Guanlin et al. 2015).

Maritime Activities, Maritime Surveillance Operations, Scanning Activities

The literature described maritime and scanning activities as the second area where the MTTSP is the most used. Strategic applications in this area of use includes the routing of security boats and vessels, as well as the scanning and classification of ships in seas and oceans, to ensure safety at both national and international level. Some examples of maritime surveillance operations are shown in Figure 3.3.1 and 3.3.2. In these examples, scanning activities conducted by marine surveillance aircraft are presented. Aside from these, other applications of the MTTSP within these areas include guiding fishing boats to areas of high fish density, and the scanning of objects with RFID devices. The current section covers studies on the application of the MTTSP in maritime and scanning activities.

In their work, Groba et al. addressed the problem of determining how fish harvesting devices would move in orbit and when the fish network would collect on the Tuna River. To solve the problem, they proposed a heuristic method based on predictive techniques, and attempted to predict the direction of movement after each time the nets were pulled out of water. Fish were defined as the targets in this study. To determine how the route optimization can be improved, a genetic algorithm based on moving estimation techniques is proposed. The solution method they proposed (GATP) was found to yield better results than the GA-TSP algorithm and NN algorithm (Carlos, Antonio and Xose 2015).

Mercer et al. have examined the maritime surveillance operation problems. They described that maritime surveillance operations are part of Australia's national security, and that, due to the movement of ships, detection every ship within range is rather difficult. In their problem, they described the aircraft to have a constant velocity v, an origin of Ø_(i), a turning radius of rc, a scanning diameter of rv, and a specific turning radius, while ships are described to have a velocity of v, a starting position of (x0,y0), and a path with a fixed angle (θ). They used the Genetic Algorithm tool in Matlab to solve the problem, and they told us that the location of the vessels and their recognizability were updated every minute.

The proposed algorithm was run for 1 second when there were 5 or 10 ships, and for 10 seconds in a wide area with 25 ships. Based on their observations, they described that minimizing the total tour length of the aircraft required the use of different scanning ranges; that narrowing the scanned area may cause certain ships to be overlooked; and that allowing the aircraft to turn with a minimum radius required an increase of 20 percent in the total route length. They also noted that having the aircraft turn with a minimum radius and distance has a significant impact on the route length (Geoff et al. 2008).

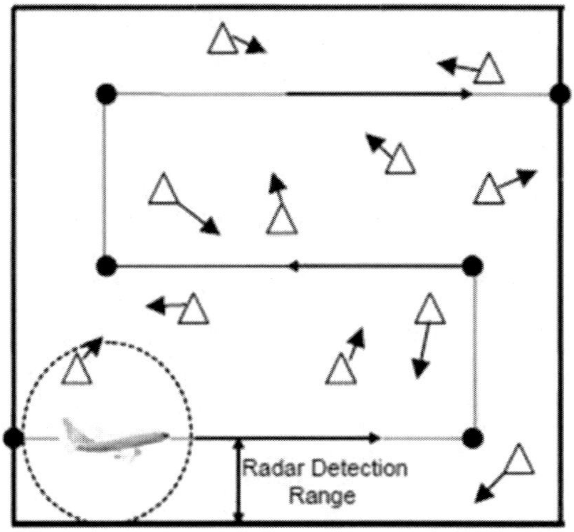

Figure 10. Maritime surveillance operation (Marlov, Kilby, Mercer 2007).

Figure 11. Maritime surveillance operation (Web5).

Killby et al. conducted a study in which they addressed a maritime surveillance problem involving one aircraft and multiple ships, and in which the velocity, direction and point of origin of the aircraft and ship were all known. Their goal was to minimize the number of unscanned ships within the mission area, and the total duration of flight. When solving the problem, they used various heuristic methods to determine the shape of the tour, and to also develop and expand the tour, which allowed them to improve the quality of the solution. They ran the problem under various scenarios, and obtained suitable routes by using the genetic algorithm in Matlab software. This study used both mobile and fixed targets, and took into account various factors such as dynamically changing data, a time window, priority restrictions, points of interception, and demand prediction. For this reason, it is considered to be more complex than the other studies in the literature (Philip et al.2007).

Marlow et al. conducted a study on a maritime surveillance problem involving a fixed mission area as well as constant aircraft velocity and scanning range, while the number of ships and their velocities were variable. When solving the problem, they compared the NN technique with the 2-opt, 2-opt stationary ships and 2-opt jumping ships heuristics, testing the algorithms under different target speeds and target numbers. Based on their findings, they determined that the 2-opt and 2-opt stationary heuristics yielded better results (Marlov, Kilby,and Mercer 2007).

Fang et al.'s study addressed a problem concerning the protection of mobile targets at sea by moving security units, based on an optimal strategy. They examined the problem based on scenarios with one-dimension and two-dimension planes, and with single or multiple security vessels. To solve the problem, they developed the CASS algorithm, and attempted to identify a suitable protection strategy that would ensure a minimum level of damage to the targets (F. Fang, A. X. Jiang, and M. Tambe 2013).

Cross et al. conducted a study on a maritime surveillance problem involving the scanning and classification of ships in any areas in the ocean waters surrounding Australia. The problem consisted of a limited area, ships moving at random speeds and directions, and a maritime surveillance aircraft monitoring this area at a certain speed and with a specific surveillance area. Their main goal was to ensure the aircraft scans and classifies all of the ships in the relevant areas with a minimum traveling time, and to identify the suitable flight path for this (Martin, David, and Jason 2007).

In a previous study, Shuttleworth et al. examined a traveling salesman problem in which, using RFID-based terminals, the water meters of houses within a certain radius were read from a certain distance by a traveling vehicle. They attempted in this problem to read the customers' water meters with a minimum traveling time or distance, and their solution method involved whole number mathematical modeling and heuristic algorithms. They tested the proposed heuristic methods on experimental data, and determined that, among five different heuristic methods, the WB4B method was simpler and more effective compared to the others (Robert et al. 2008).

Security Cameras and Surveillance Activities

Security cameras, which have become an integral part of daily life, assume a major role in many security and safety activities. One of the areas of application of the MTTSP when it comes to security cameras and surveillance units is the scanning and classification of numerous moving objects. Some of the surveillance activities carried out by the security camera is shown in Figure 12 In this context, the literature describes a number of studies concerning the application of the MTTSP to surveillance activities.

Bimbo and Pernici worked on a problem in which they attempted to have an active, optical zoom camera observe the maximum number of moving targets possible. They considered this problem as a special variation of the dynamic routing problem and kinetic traveling salesman problem, and, using their proposed method for an optimal solution, they tried to minimize the time the camera needed to observe moving targets. The effectiveness of this proposed approach was analyzed using dual camera systems, and based on their findings, Bimbo and Pernici noted that, when large number of cameras were used, this approach allowed a very wide area to be monitored at low costs (A. Del Bimbo and F. Pernici 2005).

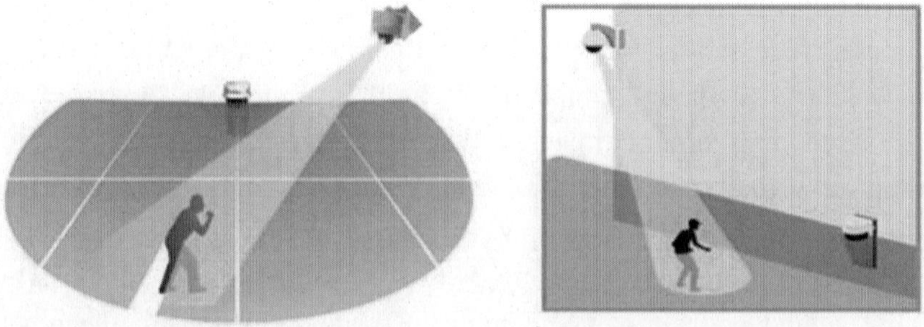

Figure 12. Scanning of moving objects with safety camera (Web 6; Web7).

In another study, Bimbo and Pernici attempted to determine a suitable surveillance/observation route for a PTZ camera that would enable a large number of moving targets to remain in the camera's field of vision. They then tested their proposed solution algorithm with the Monte Carlo Simulation, to determine the number of targets that escaped the camera's field (A.D.Bimbo and F.Pernici 2005).

Routing and Scheduling of Robotic Systems

As a result of the advances in technology, robotic systems are becoming increasingly widespread in daily life, with robots taking over many tasks previously handled by humans. Furthermore, the transition of many factories to the Industry 4.0 culture is also increasing the need for the MTTSP. Industry 4.0 that ensures activities are performed with minimum error and maximum safety and efficiency, and also reduces the time require by industrial activities. One important aspect to consider in this context is the control and guidance of robotic systems. This section covers the studies on the application of the MTTSP on robotic systems.

Ilavarasi and Joseph conducted a detailed analysis on the formulations and applications of the TSP's Beneficial Base variant, Time Window variant, Maximization Based variant and Kinetic Based variant. Taking Chalasani et al.'s previous study as a basis, Ilavarasi and Joseph described a kinetic TSP problem addressing the collection of products from conveyor belts with a k-delivery system as an application study. In addition, they also considered the MTTSP problem as the sub-type of the Kinetic TSP, and divided the problem into two parts. Both problems involved a situation where the targets move on a single line, and the number of moving targets is small. In the first of the sub-problems, a single pursuer intercepted all targets moving on a limited single line, while in the second sub-problem, some of the nodes were fixed, while the others were mobile. Problems relating to supply ships that resupply patrol boats, as well as those relating to the interception of mobile units by aircraft, are also described as being within the scope of the MTTSP (K. Ilavarasi and K. Suresh 2014).

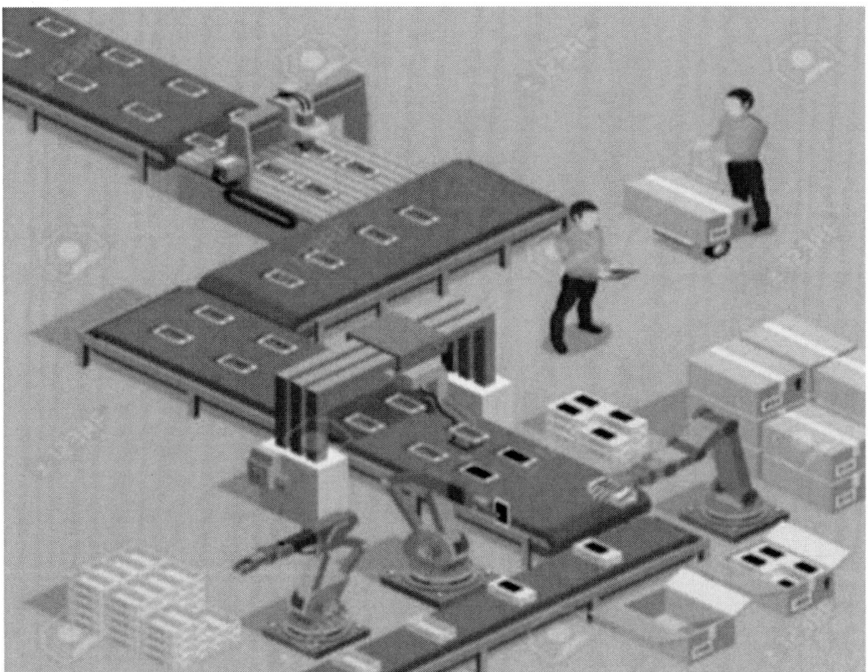

Figure 13. Example of assembly line where robotic systems are used (Web8).

Asahiro et al. conducted a study addressing the problem where n number of balls moving at a constant speed on a Euclidean plane need to be collected by k number of robots moving in a straight path from the origin; in this context, they described the MMTSP as an NP-Hard problem. The problem was examined according to three different scenarios, which were (i) the scenario where k number of robots collected n number of balls; (ii) the scenario where the goal is to determine the maximum number of balls that can be collected by k number of robots; and (iii) the scenario where the goal is to determine the minimum number of robots that can collect n number of balls. Each one of these scenarios were also evaluated according to three different variations, one in which the routes of movement are defined; a second in which these routes are not defined; and a third in which all the movement routes are the same. They also assumed that the robots moved on a path of a certain length, and that, aside from this path, they did not move or wander in any other direction. They developed an algorithm that produces solutions in a polynomial time, and provided evidence on the algorithm's validity (Yuichi et al. 2004).

In their study addressing the grasping and placing of moving objects by robots, Asahiro et al. defined a problem with an n number of products moving at constant speed on a pre-determined route, and a single robot army with the capacity to carry one object at a time. They also assumed that in each tour, the robot arm would be able to grasp and carry to the depot only a single product, and their goal in this problem was to determine the tour route resulting in the largest number of products captured/grasped.

The study also focused on the moving objects' movement on discrete lines, and the problem in question was described as an MAXSNP-Hardness type problem. To solve the problem, Asahiro et al. developed a two-factor algorithm (Yuichi, Eiji, and Shinichi 2008).

In another study, Chalasani et al. addressed a problem in which products brought by a conveyor belt are picked up by a robot with a limited transport capacity, which then carries places at a point of delivery. The problem assumed the products to be immobile or to have a known constant velocity, and the goal was to ensure the largest number of objects were grasped/picked up in the shortest period of the time. Chalasani et al. proposed an algorithm to solve the problem, which they described as an algorithm that can easily be applied to both moving and non-moving conveyor belts, and provides rapid and effective solutions (Prasad, Rajeev, and Anil 1996).

The MTTSP problem is the generic version of the TSP problem and has many features similar to those of many problems. The studies in the literature that are considered to be related to the MTTSP are shown in Table 1.

Many studies related to the MTTSP may be found in the literature, in addition to the studies shown in Table 1,2 3 and 4. It is also possible to associate the MTTSP with the scheduling and assignment problems other than the mentioned problems.

Table 1. Studies in the literature on the MTTSP (2016-2019)

Problem Type	Year			
	2016	2017	2018	2019
Dynamic Location Routing Problem	(Shangce eal. 2016)			
Dynamic TSP	(Michalis et al. 2016)	(Mohanad et al. 2017)	(Sudipta et al. 2018)	
Dynamic VRP	(Ulrike et al. 2016; R.J.Kuo et al. 2016)		(Shifeng et al. 2018; Ulmer et al. 2018)	
MTTSP		(Ukbe and Selçuk 2017)	(Groba et al. 2018)	(Ukbe and Selçuk 2019; Anke and Armin 2019; Rodrigo and Edison 2019)
Vehicle Routing Problems	(Lin et al. 2016 ; Majid et al. 2016)		(Mohammad et al. 2018)	

In the part of the study up to this point, the related problem is defined and information about the studies in the literature is given. In the next chapter, the solution methodologies in which MTTSP is used, and future studies are evaluated.

Table 2. Studies in the literature on the MTTSP (2012-2015)

	Year			
Problem Type	2012	2013	2014	2015
Dynamic TSP	(David and Sharlayne 2012; Taesu and Chelse 2012; Urszula and Lukasz; 2012)	(Anabela and Ernesto 2013; Urszula and Lukasz 2013)	(Stephen et al. 2014 ; Alejandro et al. 2014 ; Renato et al. 2014 ; Manal and Peter 2014)	(Renato 2015; Urszula and Lukasz 2015; Urszula and Lukasz 2015; Michalis et al. 2015; Michalis and Shengxiang 2015)
Dynamic VRP	(Victor 2012; Victor et al. 2012)	(Victor et al. 2013)	(Kwankaew and Paveena 2014)	(Omprakash et al. 2015)
Maritime Surveillance Problem		(Bo et al. 2013; D.O. and J.E 2013)	(Giovanni et al. 2014)	
Ship routing and Scheduling		(Marielle et al. 2013; Elin et al. 2013)	(J Mulder and R Dekker 2014; Dimitri et al. 2014)	
Technician Routing and Scheduling Problem	(Victor et al. 2012)	(V. Pillac et al. 2013)		
Traveling Salesman Problem	(M. M. Ismail et al. 2012)	(Mark et al. 2013)		(Zhouchun and Qipeng 2015)
Vehicle Routing Problems	(Ming-Che et al. 2012)		(De Jaegere et al. 2014)	(Michel et al. 2015; Gitae et al. 2015)

Table 3. Studies in the literature on the MTTSP (2007-2011)

	Year			
Problem Type	2007	2008	2009	2011
Dynamic Traveling Salesman Problem	(Gianpaolo et al. 2007)	(S. I. Sergeev 2008)	(Andrea et al. 2009; Tsung-Sheng et al. 2009)	(Anabela and Ernesto 2011; Anabela and Ernesto 2011; Weiqi 2011)
Dynamic Vehicle Routing Problems			(Shaunak et al. 2009; M. Pavone et al. 2009)	
Maritime Surveillance Problem			(L.C.C.S. Forero 2009; Marlow et al. 2009)	(Marc 2011)

Table 4. Studies in the literature on the MTTSP (2003-2006)

	Year			
Problem Type	2003	2004	2005	2006
Dynamic Traveling Salesman Problem		(Xue-Song et al. 2004; Allan et al. 2004)		(S. I. Sergeev 2006)
Maritime Surveillance Problem				(Marcel 2006)
Optimal Path Problem	(Danny et al. 2003)			
Traveling Salesman Problem		(Sándor et al. 2004)		
Vehicle Routing Problems				(Rodolfo and Jaime 2006)
Time-Dependent Orienteering Problem			(M. Knapp and Hendrik Rothe 2012)	

Table 5. Studies in the literature on the MTTSP (1995-2002)

	Year		
Problem Type	1995	2001	2002
Dynamic Traveling Salesman Problem		(Zhang et al. 2001; Michael et al. 2001)	
Maritime Surveillance Problem	K. Darby et al. 1995		
Traveling Salesman Problem			(Håkan 2002)
Time-Dependent Orienteering Problem			(Fedor and Andrzej 2002)

SOLUTION METHODS IN MTTSP

The solution of the MTTSP is very difficult and complicated, because it has a dynamic structure, the parameters in the problem change constantly, and the number of the nodes increases or decreases over time. Because of this complicated structure, the problem cannot be fully expressed through mathematical modeling. And in the studies in which mathematical model is used, time is considered as discrete. The methodologies in the literature that are used to solve the MTTSP are summarized in Figure 14.

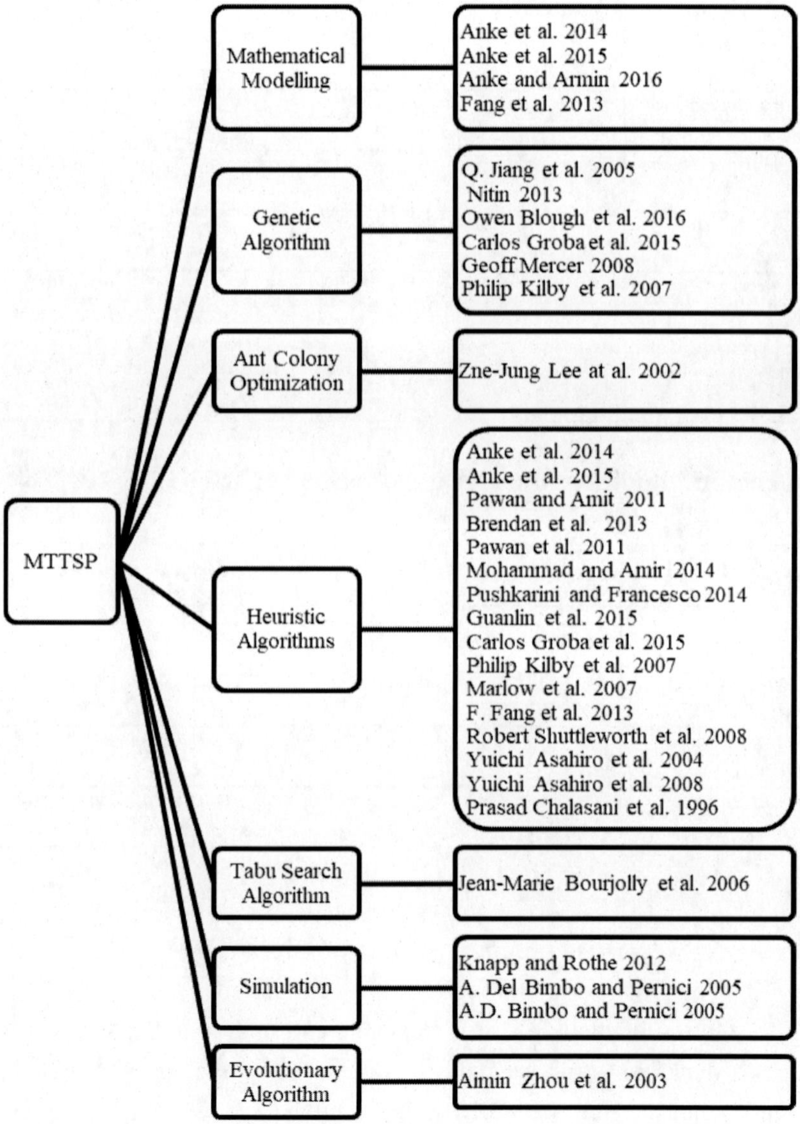

Figure 14. Solution methods in MTTSP.

When we look at Figure 14, it is seen that in solving the problem, problem-specific heuristic algorithms are used the most.

The problem has a complex structure, because it has a dynamic structure, the parameters in the problem change constantly, and the concept of time is continuous. This situation makes it difficult to produce a generic solution to the problem, and leads to the development of problem-specific heuristics. From figure 14 it is understood that heuristic or meta- heuristics methods are used in solving a large majority of the MTTSPs.

This situation basically results from the problem's dynamic structure. Mathematical model is used in very few studies in the literature, and time is considered as discrete in these studies. In addition to this, it is emphasized that, in the studies where mathematical modeling is used, as the time unit and target number increase, the time of the solution prolongs. It is not known which method is more effective, because there are limited studies on MTTSP in the literature and there is not any study comparing the mentioned methods. It is predicted that more effective solutions will be obtained by means of future studies comparing the related methods, and development of new hybrid methods.

Deficits in the Literature and Future Studies

While the MTTSP has a history as short as 20 years, it has an ever-expanding application field, and it is a kind of problem that the need for it will increase day by day as a result of technological developments. When looked at the literature, it is seen that the number of studies with respect to the problem is limited. The blind spots as well as the implications for the future are explained below.

1) In the studies included in the literature, generally problem-specific heuristics are used. The suggested heuristic algorithms can produce solution to the related problem, but it may become inadequate when the problem structure changes slightly. In addition, it is determined from the conducted studies that as the

number of targets, the number of strikers, and the time of the task in the problem increase, the solution time increases significantly. It is predicted that exact and heuristic algorithms that can produce more efficient solutions in a shorter time and can be applied easily to different variants of the problem may be developed in future studies.

2) In the literature, coordinate axes are generally considered as having two dimensions (x and y axis); when we look at the present technology, solution methodologies taking into consideration three dimensions (x, y and z axis) for the systems like unmanned aircrafts and drones are needed. It is thought that MTTSP studies taking into consideration the three dimensions will increase in the future.
3) The Resupply situation is taken into account very little in the studies included in the literature. The systems such as armed unmanned aircrafts and drones have a certain ammunition or fuel capacity and they need to be refueled within the relevant task period. It is estimated that future studies that take into consideration the matters mentioned will increase.
4) The stochasticity is taken into account very little in the studies included in the literature. However, many tasks, especially military operations, contain stochastic processes. Issues such as what the number of the targets will be, in what orbit the targets will move, and how long the task will last are the examples of the stochastic processes. It is foreseen that in the coming years, the subject of stochastic will be more examined under the MTTSP.
5) In the studies included in the literature, it is tried to destroy the targets, included in the system, by means of certain number of strikers, within the scope of MTTSP. When we look at military operations, it is known that the vehicles and ammunition used in the related operations have high costs. In this context, it is thought that within the MTTSP the targets are required to be destroyed with minimum number of vehicles and by taking into account the task period as well as the task cost. In addition, as a result of the

literature review, it is determined that there is no study taking into consideration the mentioned matters within the MTTSP.

6) In the studies in the literature, generally the situations, in which the targets run and the strikers pursue, are taken into consideration. In real life problems, it is seen that the targets are armed, too, and according to the power of these weapons the targets can inflict damage to the striker. Besides, the MTTSP is a kind of problem that should not considered only as an attack-oriented but also as a defense-oriented problem. In the problem, for instance, let a target set consisting of S = {S1, S2, ..., Sn} have had an attack on a set of strikers consisting of P = {P1, P2 ... Pn}. Let inside the system there be military bases such as headquarters having the set of K = {K1, K2, ..., Kn}, data processing center and ammunition depots, which belongs to the strikers. Under these assumptions, it is thought that the problem of destroying all targets with minimum loss within minimum task period will arise and it is envisaged that the problem will be named as defense-oriented MTTSP. In addition, the mentioned problem, as far as it is known, has been expressed for the first time in the literature.
7) While search and rescue operations are, according to the literature, an area which has not been extensively studied and they form one of the most humanitarian area of application of MTTSP. Advances in unmanned aerial vehicles and similar equipment are enabling safer and more rapid intervention in disaster zones. Possible areas of application for the MTTSP includes, for example, the rescue of multiple persons or vehicles caught in a flood zone, or the rapid detection of numerous persons in search-and-rescue operations conducted on a mountain or at sea. In addition to these, the tracking, capture or destruction of vehicles belonging to suspects or criminals, including bomb-laden vehicles, is another area that can be considered within the scope of the MTTSP's application areas.
8) With the rapid advances in technology, unmanned machine systems are becoming more widely used with each passing day,

with robotic systems finding use in many areas from defense to manufacturing, healthcare and education. Manufacturing in one of the areas where robotic systems are finding the most use. These systems allow heavy, difficult and repetitive tasks to be executed more easily and safely than they would be by humans, and with a minimum level of error, which is why their use in the industry is becoming increasingly widespread. As the Industry 4.0 concept begins to make its way into manufacturing systems, it is believed that this process will further accelerate the growing use of robotic systems. The increasing use of robotic systems is likely to be accompanied by the problem of proper routing and scheduling to ensure maximum efficiency at minimum cost. It is believed that the mobility of robotic systems and products, as well as the emphasis on minimum cost and maximum efficiency, will lead to a greater use of the MTTSP in the coming years to address this type of problem.

9) In the present-day world, many countries are faced with the threat of terror attacks, and have to take a broad range of measures to ensure their national security. These measures can be against migrant activities in the seas or mobile terrorist units. In parallel with advances in technology, the majority of these measures rely on modern weapons and communication technologies. In this context, use of MTTSP is predicted for guiding and controlling these weapon technologies, while also enabling them to complete their missions at minimum cost. Especially because unmanned aerial vehicles are widespread and used in many tasks, these vehicles must be effectively routed and schedules. Unmanned aerial vehicles are thought to be frequently evaluated in MTTSP in the years to come. It is believed that missile defense systems will become one of the major areas in which this problem will be employed.
10) Security cameras are widely used in many places such as houses, airports, stadiums, shopping malls, hospitals. In the coming years,

one of the areas where the MTTSP is likely to find the most use is the scanning and classification activities of security cameras.

The identified deficits regarding the MTTSP and the evaluations relating to future studies are mentioned above. It is possible to increase the number of these items; it is predicted that the problem will have a wider application area if it is examined from different viewpoints.

CONCLUSION

While the Moving Target Traveling Salesman Problem represents a general situation of the Traveling Salesman Problem, it also differs from this problem in that the locations of cities or nodes varies over time. The problem assumes one pursuer and n number of moving targets, and that the velocity and locations of the pursuer and targets are known. It also attempts to determine the route that will result in the interception of all targets under certain conditions or limitations.

The MTTSP now has a wide application area ranging from military operations to arrangement of the conveyor belts, from scheduling of robotic systems to routing of satellites. With technologies developing, this application area is getting wider and wider and the need for the problems is getting bigger and bigger.

One of the areas in which the problem is applied the most is comprised by military operations. The unmanned aircraft and drones being armed, being fitted with systems of high-level vision, and being utilized in many areas, especially in search and rescue operations, lead the countries to establish unmanned armies. The efficient routing of, and controlling, these vehicles pose great importance. The MTTSP is frequently used in solving the specified problem, and has the purpose of proposing solution methods in which all targets are destroyed within minimum task period. In addition, marine surveillance operations, positioning of satellites, scheduling of the robotic systems, directing the conveyor lines, and routing the security cameras constitute other application areas of the problem. Besides, it is also predicted that the problem will be frequently used in search and rescue activities in the coming years.

In this study, the studies in the literature related to the MTTSP have been examined in detail. First, the problem has been defined, and some of its variants have been expressed. In the following chapters, on the basis of the studies contained in the literature, the problem has been classified according to its application areas, and the methodologies used in the solution of the problem have been described. Finally, evaluations have been made with respect to the deficits found in the literature and regarding the studies in the future. In addition, it is thought that, as far as it is known, this study has been the first survey about the MTTSP. It is estimated that with the improvements in artificial intelligence technologies, the semiautomatic and automatic systems will be used more intensively in many areas, and therefore, it is thought that the interest in the problem will increase even further. Besides these, it is foreseen that with the development of new solution methods and the use of hybrid algorithms, more efficient solutions in shorter time periods can be achieved for large size problems.

ACKNOWLEDGMENT

This work is supported by the Tubitak-BIDEB 2211 PhD Scholarship Program.

REFERENCES

Adham Manal T., and Peter J. Bentley. (2014, December). "An Artificial Ecosystem Algorithm Applied to Static and Dynamic Travelling Salesman Problems." In *Evolvable Systems (ICES), 2014 IEEE International Conference on – IEEE*: 149-156.

Agharkar Pushkarini, and Francesco Bullo. 2014. "Vehicle Routing Algorithms to Intercept Escaping Targets," *In American Control Conference (ACC) - IEEE*: 952-957.

Akandwanaho Stephen M., Aderemi O. Adewumi, and Ayodele A. Adebiyi. 2014. "Solving Dynamic Traveling Salesman Problem Using Dynamic Gaussian Process Regression." *Journal of Applied Mathematics*.

An Bo, Fernando Ordóñez, Milind Tambe, Eric Shieh, Rong Yang, Craig Baldwin, Joseph DiRenzo III, Kathryn Moretti, Ben Maule, and Garrett Meyer. 2013. "A Deployed Quantal Response-Based Patrol Planning System for the US Coast Guard." *Interfaces* 43(5): 400-420.

Asahiro Yuichi, Eiji Miyano, and Shinichi Shimoirisa. 2008. "Grasp and delivery for moving objects on broken lines." *Theory of Computing Systems* 42(3): 289-305.

Asahiro Yuichi, Takashi Horiyama, Kazuhisa Makino, Hirotaka Ono, Toshinori Sakuma, and Masafumi Yamashita. 2004. "How to Collect Balls Moving in the Euclidean Plane." *Electronic Notes in Theoretical Computer Science* 91, 229-245.

Bimbo A. D., and F. Pernici. (2005, June) "Saccades planning with kinetic TSP for distant targets identification. In Imaging for Crime Detection and Prevention." *2005. ICDP 2005. The IEE International Symposium on – IET:* 145-149.

Bimbo A. Del, and F. Pernici. (2005, October) "Distant Targets Identification As an On-Line Dynamic Vehicle Routing Problem Using an Active-Zooming Camera." *In Visual Surveillance and Performance Evaluation of Tracking and Surveillance, 2005. 2nd Joint IEEE International Workshop on - IEEE*: 97-104.

Blough Owen P., Tyler K. Farrington, and Jennifer Hudson. 2016. "Trojan Asteroid Mission Design: Target Selection And Sequencing Optimization." Conference: 26th AAS/AIAA Space Flight Mechanics Meeting, At Napa-CA.

Bopardikar Shaunak D., Stephen L. Smith, Francesco Bullo, and Joao P. Hespanha. (2009, June) "Dynamic Vehicle Routing with Moving

Demands-Part I: Low Speed Demands and High Arrival Rates." *In American Control Conference, 2009. ACC'09.- IEEE*: 1454-1459.

Boryczka Urszula, and Łukasz Strąk. (2012, November). "A Hybrid Discrete Particle Swarm Optimization with Pheromone for Dynamic Traveling Salesman Problem." *In International Conference on Computational Collective Intelligence - Springer Berlin Heidelberg*: 503-512.

Boryczka Urszula, and Łukasz Strąk. (2013, September). "Efficient DPSO Neighbourhood for Dynamic Traveling Salesman Problem." *In International Conference on Computational Collective Intelligence - Springer Berlin Heidelberg*: 721-730.

Boryczka Urszula, and Łukasz Strąk. (2015, March). "Diversification and Entropy Improvement on the DPSO Algorithm for DTSP., *In Asian Conference on Intelligent Information and Database Systems - Springer International Publishing:* 337-347.

Boryczka Urszula, and Łukasz Strąk. 2015. "Heterogeneous DPSO Algorithm for DTSP." *In Computational Collective Intelligence - Springer International Publishing:* 119-128.

Bourjolly Jean-Marie, Ozgur Gurtuna, and Aleksander Lyngvi. 2006. "On-Orbit Servicing: A Time-Dependent, Moving-Target Traveling Salesman Problem." *International Transactions in Operational Research* 13(5): 461-481.

Chalasani Prasad, Rajeev Motwani, and Anil Rao. (1996, July). "Algorithms for robot grasp and delivery." *In 2nd International Workshop on Algorithmic Foundations of Robotics.*

Chang Tsung-Sheng, Yat-wah Wan, and Wei Tsang OOI. 2009. "A Stochastic Dynamic Traveling Salesman Problem with Hard Time Windows." *European Journal of Operational Research* 198(3): 748-759.

Chang Tsung-Sheng, Yat-wah Wana, and Wei Tsang OOI. 2009. "A Stochastic Dynamic Traveling Salesman Problem with Hard Time Windows." *European Journal of Operational Research* 198(3): 748-759.

Chen Lin, Shan Lin, and Hua Huang. (2016, July). "Charge Me If You Can: Charging Path Optimization and Scheduling in Mobile Networks," *In Proceedings of the 17th ACM International Symposium on Mobile Ad Hoc Networking and Computing – ACM:* 101-110.

Chen Shifeng, Rong Chen, Gai-Ge Wang, Jian Gao, and Arun Kumar Sangaiah. 2018. "An Adaptive Large Neighborhood Search Heuristic for Dynamic Vehicle Routing Problems." *Computers & Electrical Engineering* 67: 596-607.

Chena Danny Z., Ovidiu Daescu, Xiaobo (Sharon) Hu, and Jinhui Xu. 2003. "Finding an optimal path without growing the tree." *Journal of Algorithms* 49 (1): 13-41.

Cheong Taesu, and Chelsea C. White. 2012. "Dynamic Traveling Salesman Problem: Value of Real-Time Traffic Information." *IEEE Transactions on Intelligent Transportation Systems* 13(2): 619-630.

Choi In-Chan, Seong-In Kim, and Hak-Soo Kim. 2003. "A Genetic Algorithm with A Mixed Region Search for the Asymmetric Traveling Salesman Problem." *Computers & Operations Research* 30 (5): 773-786.

Choubey Nitin S., 2013. "Moving Target Travelling Salesman Problem Using Genetic Algorithm." *International Journal of Computer Applications* 70(2): 30-34.

Chowdhury Sudipta, Mohammad Marufuzzaman, Huseyin Tunc, Linkan Bian, and William Bullington. 2018. "A Modified Ant Colony Optimization Algorithm to Solve A Dynamic Traveling Salesman Problem: A Case Study with Drones for Wildlife Surveillance." *Journal of Computational Design and Engineering*.

Christiansen Marielle, Kjetil Fagerholt, Bjørn Nygreen, and David Ronen. 2013. "Ship Routing and Scheduling in the New Millennium." *European Journal of Operational Research* 228(3): 467 – 483.

Cross Martin, David Marlow and Jason Looker. 2007 "Application of the non-stationary travelling salesman problem to maritime surveillance." *Proceedings of MISG*:1-4.

Darby-Dowman K., R. K. Fink, G. Mitra, and J. W. Smithc. 1995. "An Intelligent System for US Coast Guard Cutter Scheduling." *European Journal of Operational Research* 87(3): 574-585.

De Jaegere, N., Defraeye, M., & Van Nieuwenhuyse, I. 2014. "The Vehicle Routing Problem: State of the Art Classification and Review."

de Moraes Rodrigo S., and Edison P. de Freitas. 2019. "Experimental Analysis of Heuristic Solutions for the Moving Target Traveling Salesman Problem Applied to a Moving Targets Monitoring System." *Expert Systems with Applications*.

Dondo Rodolfo, and Jaime Cerdá. 2006. "A Reactive MILP Approach to the Multidepot Heterogeneous Fleet Vehicle Routing Problem with Time Windows," *International Transactions in Operational Research* 13(5): 441-459.

Englot Brendan, Tuhin Sahai, and Isaac Cohen. 2013. "Efficient Tracking and Pursuit of Moving Targets by Heuristic Solution of The Traveling Salesman Problem." In *Decision and Control (CDC),* 2013 IEEE 52nd Annual Conference on: 3433-3438.

Fajardo David, and Sharlayne Waller. 2012. "Dynamic Traveling Salesman Problem in Stochastic-State Network Setting for Search-And-Rescue Application." *Transportation Research Record: Journal of the Transportation Research Board* 2283:122-130.

Fang F., A. X. Jiang and M. Tambe. 2013. "Protecting moving targets with multiple mobile resources." *Journal of Artificial Intelligence Research* 48: 583-634.

Fekete Sándor P., Rudolf Fleischer, Aviezri Fraenkel, and Matthias Schmittd. 2004. "Traveling Salesmen in the Presence of Competition," *Theoretical Computer Science* 313(3): 377-392.

Fomina Fedor V., and Andrzej Lingas. 2002. "Approximation Algorithms for Time-Dependent Orienteering." *Information Processing Letters* 83(2): 57-62.

Forero L. C. C. S., (2009, January). "Development of a Model to Optimize Maritime Surveillance and Patrol Operations in Colombian Navy (Presentation Supporting Paper)." *In IIE Annual Conference. Proceedings (p. 1) -Institute of Industrial Engineers-Publisher.*

Gao Shangce, Yirui Wang, Jiujun Cheng, Yasuhiro Inazumi, and Zheng Tang. 2016. "Ant colony optimization with clustering for solving the dynamic location routing problem." *Applied Mathematics and Computation* 285: 149-173.

Gendreau Michel, Gianpaolo Ghiani, and Emanuela Guerriero. 2015. "Time-dependent routing problems: A review." *Computers & Operations Research* 64: 189-197.

Ghiani Gianpaolo, Antonella Quaranta and Chefi Triki. 2007. "New Policies for the Dynamic Traveling Salesman Problem." *Optimisation Methods and Software* 22 (6): 971-983.

Goyal Sanchit. 2010. "A Survey on Travelling Salesman Problem." *In Midwest Instruction and Computing Symposium*: 1-9.

Grob Marcel J. H. B., 2006. "Routing of Platforms in A Maritime Surface Surveillance Operation." *European Journal of Operational Research* 170(2): 613-628.

Groba Carlos, Antonio Sartal, and Xosé H. Vázqueza. 2015. "Solving The Dynamic Traveling Salesman Problem Using A Genetic Algorithm with Trajectory Prediction: An Application to Fish Aggregating Devices." *Computers & Operations Research* 56: 22-32.

Groba Carlos, Antonio Sartal, and Xosé H.Vázquez. 2018. "Integrating Forecasting in Metaheuristic Methods to Solve Dynamic Routing Problems: Evidence From the Logistic Processes of Tuna Vessels." *Engineering Applications of Artificial Intelligence* 76: 55-66.

Guntsch Michael, Martin Middendorf, and Hartmut Schmeck. (2001, July). "An Ant Colony Optimization Approach to Dynamic TSP." *In Proceedings of the 3rd Annual Conference on Genetic and Evolutionary Computation - Morgan Kaufmann Publishers Inc. 860-867.*

Halvorsen-Weare Elin E., Kjetil Fagerholt, and Mikael Rönnqvist. 2013. "Vessel Routing and Scheduling Under Uncertainty in the Liquefied Natural Gas Business," *Computers & Industrial Engineering* 64(1): 290-301.

Hammar Mikael, and Bengt J. Nilsson. (1999, July). "Approximation Results for Kinetic Variants of TSP." *In International Colloquium on*

Automata, Languages, and Programming -Springer Berlin Heidelberg: 392-401.

Hammar Mikael, and Bengt J. Nilsson. 2002. "Approximation results for kinetic variants of TSP." *Discrete & Computational Geometry* 27 (4): 635-651.

Hammar Mikael, and Bengt J. Nilsson. 2002. "Approximation Results for Kinetic Variants of TSP." *Discrete & Computational Geometry* 4(27).

Helvig C. S., Gabriel Robins, and Alex Zelikovsky. (1998, August). "Moving-Target TSP and Related Problems." In *European Symposium on Algorithms*: 453-464.

Helvig C. S., Gabriel Robins, and Alex Zelikovsky. 2003. "The Moving-Target Traveling Salesman Problem." *Journal of Algorithms* 49 (1): 153-174.

Huang Zhang-Can, Xiao-Lin Hu, and Si-Duo Chen. 2001. "Dynamic Traveling Salesman Problem Based on Evolutionary Computation." *In Evolutionary Computation, 2001. Proceedings of the 2001 Congress on - IEEE.* 2: 1283-1288.

Huang Zhouchun, and Qipeng P. Zheng. 2015. "Decomposition-Based Exact Algorithms for Risk-Constrained Traveling Salesman Problems with Discrete Random Arc Costs," *Optimization Letters* 9(8): 1553-1568.

Ilavarasi K., and K. Suresh Joseph. (2014, February). "Variants of Travelling Salesman Problem: A Survey." *In Information Communication and Embedded Systems (ICICES), 2014 International Conference on* – IEEE:1-7.

Ismail M. M., Muhammad Iqbal Zakaria, Amar Faiz Zainal Abidin, Juwita Mad Juliani, Asrani Lit, S. Mohammad Mirjalili, Nur Anis Nordin, and Muhammad Faiz Mohamed Saaid. 2012. "Magnetic Optimization Algorithm Approach for Travelling Salesman Problem." *World Academy of Science, Engineering and Technology* 62: 1393-1397.

Jabbarpour Mohammad Reza, Houman Zarrabi, Rashid Hafeez Khokhar, Shahaboddin Shamshirband, and Kim-Kwang Raymond Choo. 2018. "Applications of Computational Intelligence in Vehicle Traffic Congestion Problem: A Survey," *Soft Computing* 22(7): 2299–2320.

Jiang Q., R. Sarker and H. Abbass. 2005. "Tracking Moving Targets And The Non-Stationary Traveling Salesman Problem," *Complexity International* 11(2005): 171-179.

Jindal Pawan, Amit Kumar, and Shishir Kumar. 2011. "Dynamic Version of Traveling Salesman Problem," *International Journal of Computer Applications (0975–8887)* 19(1): 37- 46.

Jindal Pawan, and Amit Kumar. 2011. "Multiple Target Intercepting Traveling Salesman Problem," *International Journal of Computer Science and Technology* 2(2): 327-331.

John Jurjee Mohanad Muayad, Hafiz Mohd Sarim, Noora Hani Abdulmajeed Al-Dabbagh, And Erna Budhiarti Nababan. 2017. "A Multi-Population Harmony Search Algorithm For The Dynamic Travelling Salesman Problem With Traffic Factors." *Journal of Theoretical and Applied Information Technology* 95(2).

Jonsson Håkan, 2002. "The Traveling Salesman Problem for Lines in the Plane," *Information Processing Letters* 82(3): 137-142.

Kaiwartya Omprakash, Sushil Kumar, D. K. Lobiyal, Pawan Kumar Tiwari, Abdul Hanan Abdullah, and Ahmed Nazar Hassan. 2015. "Multiobjective Dynamic Vehicle Routing Problem And Time Seed Based Solution Using Particle Swarm Optimization." *Journal of Sensors* (2015).

Khosravi Mohammad, and Amir G. Aghdam. (2014, December). "Cooperative Receding Horizon Control for Multi-Target Interception in Uncertain Environments." In *Decision and Control (CDC), 2014 IEEE 53rd Annual Conference on:* 4497-4502.

Kilby Philip, Patrick Tobin, Ruth Luscombe, Steven I. Barry, and Roslyn Hickson. 2007. "The Maritime Surveillance Problem."

Kim Gitae, Yew-Soon Ong, Chen Kim Heng, Puay Siew Tan, and Nengsheng Allan Zhang. 2015. "City Vehicle Routing Problem (City VRP): A Review," *IEEE Transactions on Intelligent Transportation Systems* 16(4): 1654-1666.

Knapp M., and Hendrik Rothe. 2012. "Concept for Simulating Engagement Strategies for C-RAM Systems Using Laser Weapons." *Proceedings of the DMMS*, (2012).

Kuo R. J., B. S. Wibowo, and F. E. Zulviaa. 2016. "Application of A Fuzzy Ant Colony System to Solve The Dynamic Vehicle Routing Problem with Uncertain Service Time." *Applied Mathematical Modelling* 40 (23): 9990-10001.

Lai Ming-Che, Hansuk Sohn, Bill Tseng, and Dennis L. Bricker. 2012. "A Hybrid Benders/Genetic Algorithm for Vehicle Routing and Scheduling Problem." *International Journal of Industrial Engineering* 19(1): 33-46.

Larsen Allan, Oli B.G. Madsen, and Marius M. Solomon. 2004. "The A Priori Dynamic Traveling Salesman Problem with Time Windows." *Transportation Science* 38(4): 459-472.

Lexter Mark, David De Lara, B Burgos, Silva, and Allen Nazareno. 2013. "Determining the Optimal Route of Vehicles Delivering Relief Goods to the Calamity-Prone Areas in Region," *Journal of Nature Studies* 13(2): 13-24.

Marlow D. O., and J. E. Murphy. (2013, December). "Testing Various Backtracking Algorithms in Airborne Maritime Surveillance Modelling," In *20th International Congress on Modelling and Simulation, Adelaide, Australia* 1(6).

Marlow, D. O., P. Kilby, and G. N. Mercer. (2009, July). "Examining Methods for Maximizing Ship Classifications in Maritime Surveillance." *In 18th World IMACS/MODSIM Congress, Cairns, Australia*: 1630-1632.

Marlow, D. O., P. Kilby, and G. N. Mercer. 2007. "The Travelling Salesman Problem in Maritime Surveillance–Techniques, Algorithms and Analysis." *In Proceedings of the International Congress on Modelling and Simulation*: 684-690.

Mavrovouniotis Michalis, and Shengxiang Yang. (2015, April). "Applying Ant Colony Optimization To Dynamic Binary-Encoded Problems." *In European Conference on the Applications of Evolutionary Computation - Springer International Publishing*: 845-856.

Mavrovouniotis Michalis, Felipe M Müller and Shengxiang Yang. (2015, July). "An Ant Colony Optimization Based Memetic Algorithm for the Dynamic Travelling Salesman Problem." *In Proceedings of the 2015*

Annual Conference on Genetic and Evolutionary Computation - ACM: 49-56.

Mavrovouniotis Michalis, Felipe M. Müller, and Shengxiang Yang. 2016. "Ant Colony Optimization With Local Search for Dynamic Traveling Salesman Problems." *IEEE Transactions on Cybernetics.*

Meesuptaweekoon Kwankaew, and Paveena Chaovalitwongse. 2014. "Dynamic Vehicle Routing Problem With Multiple Depots." *Engineering Journal* 18(4): 135-149.

Mei Guanlin, Xiaomin Ran, Deliang Fang, and Ce Zhang. 2015. "Improved Satellite Scheduling Algorithm for Moving Target." *In Proceedings of The fourth International Conference on Information Science and Cloud Computing (ISCC2015), 18-19 December 2015. Guangzhou, China.*

Mercer Geoff, Steven I. Barry, David O. Marlow, and Philip Kilby. 2008. "Investigating the Effect of Detection and Classification Range and Aircraft Dynamics On A," *ANZIAM Journal* 49: 475-492.

Mulder J, and R. Dekker. 2014. Methods for Strategic Liner Shipping Network Design. *European Journal of Operational Research* 235(2): 367- 377.

Pantrigo Juan José, and Abraham Duarte. 2013. "Low-Level Hybridization of Scatter Search and Particle Filter for Dynamic TSP Solving." *Metaheuristics for Dynamic Optimization*: 291-308.

Pantuso Giovanni, Kjetil Fagerholt, and Lars Magnus Hvattuma. 2014. "A Survey On Maritime Fleet Size And Mix Problems." *European Journal of Operational Research* 235(2): 341-349.

Papageorgiou Dimitri J., George L. Nemhauser, Joel Sokol, Myun-Seok Cheon, and Ahmet B. Keha. 2014. "MIRPLib–A Library Of Maritime Inventory Routing Problem Instances: Survey, Core Model, and Benchmark Results." *European Journal of Operational Research* 23(2): 350-366.

Pavone M., N. Bisnik, E. Frazzoli, and V. Isler. 2009. "A Stochastic And Dynamic Vehicle Routing Problem With Time Windows And Customer Impatience," *Mobile Networks and Applications* 14(3): 350-364.

Pillac V., C. Guéret, and A. L. Medaglia. 2013. "A Parallel Matheuristic for the Technician Routing and Scheduling Problem," *Optimization Letters* 7(7): 1525 – 1535.

Pillac Victor, 2012. "Dynamic Vehicle Routing: Solution Methods and Computational Tools," Doctoral Dissertation, Ecole des Mines de Nantes.

Pillac Victor, Christelle Guéret, and Andrés Medaglia. 2012. "On the Dynamic Technician Routing and Scheduling Problem," *Conference: Proceedings of the 5th International Workshop on Freight Transportation and Logistics (ODYSSEUS 2012).*

Pillac Victor, Christelle Guéreta, and Andrés L.Medagliab. 2012. "An Event-Driven Optimization Framework for Dynamic Vehicle Routing." *Decision Support Systems* 54 (1): 414-423.

Pillac Victor, Michel Gendreau, Christelle Guéreta, and Andrés L. Medagliab. 2013. "A Review of Dynamic Vehicle Routing Problems." *European Journal of Operational Research* 225(1): 1-11.

Raja Balachandar S., and K. Kannan. 2007. "Randomized Gravitational Emulation Search Algorithm for Symmetric Traveling Salesman Problem." *Applied Mathematics and Computation* 192 (2): 413-421.

Ritzinger Ulrike, Jakob Puchinger, and Richard F. Hartl. 2016. "A Survey on Dynamic and Stochastic Vehicle Routing Problems," *International Journal of Production Research* 54 (1): 215-231.

Sergeev S. I., 2006. "Discrete optimization by optimal control methods. III. The Dynamic Traveling Salesman Problem." *Automation and Remote Control* 67 (7): 1039-1050.

Sergeev S. I., 2008. "Hybrid Control Systems and The Dynamic Traveling Salesman Problem," *Automation and Remote Control* 69(1): 42-51.

Shuttleworth Robert, Bruce L. Golden, Susan Smith, and Edward Wasil. 2008. "Advances In Meter Reading: Heuristic Solution of The Close Enough Traveling Salesman Problem over A Street Network," *In The Vehicle Routing Problem: Latest Advances and New Challenges:* 487-50.

Simões Anabela, and Ernesto Costa. (2011, April). "CHC-Based Algorithms for the Dynamic Traveling Salesman Problem." *In*

European Conference on the Applications of Evolutionary Computation- Springer Berlin Heidelberg: 354-363.

Simões Anabela, and Ernesto Costa. (2011, July). "Memory-Based CHC Algorithms for the Dynamic Traveling Salesman Problem." *In Proceedings of the 13th Annual Conference on Genetic and Evolutionary Computation – ACM*: 1037-1044.

Simões Anabela, and Ernesto Costa. (2013, July). "Extended Virtual Loser Genetic Algorithm for the Dynamic Traveling Salesman Problem," *In Proceedings of the 15th annual conference on Genetic and evolutionary computation- ACM*: 869-876.

Smith Stephen L., Shaunak D. Bopardikar, and Francesco Bullo. 2009. "A Dynamic Boundary Guarding Problem with Translating Targets." *In Decision and Control, 2009 held jointly with the 2009 28th Chinese Control Conference. CDC/CCC 2009. Proceedings of the 48th IEEE Conference on*: 8543-854).

Stieber Anke, and Armin Fügenschuh. 2016. "Variants in Modeling Time Aspects for the Multiple Traveling Salesmen Problem with Moving Targets." *Helmut-Schmidt-Universität, Universität der Bundeswehr Hamburg, Fachbereich Maschinenbau.*

Stieber Anke, and Armin Fügenschuh. 2018. "The Multiple Traveling Salesmen Problem with Moving Targets and Nonlinear Trajectories." In *Operations Research Proceedings 2017 - Springer, Cham. (pp. 489-494).*

Stieber Anke, and Armin Fügenschuh. 2019. "Dealing with Time in the Multiple Traveling Salesmen Problem with Moving Targets."

Stieber Anke, Armin Fügenschuh, Maria Epp, Matthias Knapp, and Hendrik Rothe. 2014. *"The Multiple Traveling Salesmen Problem with Moving Targets."* Helmut-Schmidt-Univ., Professur für Angewandte Mathematik.

Stieber Anke, Armin Fügenschuh, Maria Epp, Matthias Knapp, and Hendrik Rothe. 2015. "The Multiple Traveling Salesmen Problem with Moving Targets." *Optimization Letters* 9 (8): 1569 -1583.

Tinós Renato, Darrell Whitley and Adele Howe. (2014, July)."Use of Explicit Memory in The Dynamic Traveling Salesman Problem." *In*

Proceedings of the 2014 Annual Conference on Genetic and Evolutionary Computation - ACM: 999-1006.

Tinós Renato. 2015. "Analysis of the Dynamic Traveling Salesman Problem with Weight Changes." *In Computational Intelligence (LA-CCI), 2015 Latin America Congress on – IEEE:* 1-6.

Toriello Alejandro, William B. Haskell, and Michael Poremba. 2014. "A Dynamic Traveling Salesman Problem with Stochastic Arc Costs." *Operations Research* 62(5): 1107-1125.

Uçar Ukbe Üsame, and Selçuk Kürşat İşleyen. 2017. "A Solution Approach Based on Simulated Annealing for the Destruction of Moving Targets in Time Window by Air Operations," *8th InternationalAdvanced Technologies Symposium (IATS).*

Uçar Ukbe Üsame, and Selçuk Kürşat İşleyen. 2019. "A New Solution Approach for UAV Routing Problem with Moving Target – Heterogeneous Fleet." *Journal of Polytechnic.* DOI: 10.2339/politeknik.466393.

Ulmer, M. W., Soeffker, N., & Mattfeld, D. C. (2018). Value Function Approximation for Dynamic Multi-Period Vehicle Routing. European Journal of Operational Research 269(3),:883-899.

Varga Andrea, Camelia Chira, and Dan Dumitrescu. (2009, April). "A Multi-agent Approach to Solving Dynamic Traveling Salesman Problem." *In E. Calin, B. L. Iantovics, & F. G. Filip (Eds.), AIP Conference Proceedings - AIP* 1117 (1): 189-197.

Web1.https://www.google.com.tr/search?q=Unmanned+aerial+vehicles+loaded+with+ammunition&rlz=1C1NDCM_trTR717TR717&source=lnms&tbm=isch&sa=X&ved=0ahUKEwj3gsaR-fbVAhWEYlAKHYRwBc4Q_AUICygC&biw=1517&bih=681#imgrc=wv4fU9BAYzpv1M: Accessed 30 August 2017.

Web2.https://www.google.com.tr/search?q=lazer+silahlar%C4%B1+ile+savunma&tbm=isch&tbs=rimg:CYPT4dN7HhIjIjg_1c0BIouOk2HJQCgQu1vwHCljv3joc0xQHc3VlmFIk0Xr3uM0MVotdtjQrAL1Vdj3Dm5S92hQdOSoSCT9zQEii46TYEUwJvgu0mjHNKhIJclAKBC7W_1AcR3TaE35qtIEAqEgkKWO_1eOhzTFBGf2X6ZGbQDGioSCQdzdWWYUiTREfK7bvgPq9Y_1KhIJeve4zQxWi10RxdAOnpKkUigqEgm2N

CsAvVV2PRG1JQc4Z9SRCCoSCcOblL3aFB05EWI1DxS4qjfM&tbo=u&sa=X&ved=0ahUKEwiQyMaB9fbVAhVNZ1AKHZTVCBcQ9C8IHA&biw=1517&bih=681&dpr=0.9#imgrc=HO_ndFRlbEnCtM: Accessed 30 August 2017.

Web3.https://dabrownstein.com/2016/11/25/we-think-our-shores-are-stable-but-they-are-not/ Accessed 30 August 2017.

Web4.https://www.google.com.tr/search?biw=1517&bih=735&tbm=isch&sa=1&q=reFueling+the+satellites+animation&oq=reFueling+the+satellites+animation&gs_l=psyab.3...3591.8301.0.8709.12.12.0.0.0.0.123.1222.0j11.11.0....0...1.1.64.psy-ab..1.0.0.U883B28RxzQ#imgrc=f9rAJgNkyCp1sM: Accessed 30 August 2017.

Web5.https://www.youtube.com/watch?v=IUXw2tYbJZk Accessed 30 August 2017.

Web6.https://www.google.com.tr/search?q=Scan+moving+object+with+safety+camera&tbm=isch&tbs=rimg:CT62_1ywabhGpIjhZLZfUP2jwBrcyDhvajMoYWVB6ngvVYj3C9rc6aLIfczGKDL-OhT5OBY-vSKx5_1LqzJb8CyafDVioSCVktl9Q_1aPAGESB5wz6bi6RLKhIJtzIOG9qMyhgR-HI440I0jzwqEglZUHqeC9ViPRGEZYsRRNt0PyoSCcL2tzposh9zEbRRjNd3efRGKhIJMYoMv46FPk4REiUfwchoyR8qEgkFj69IrHn8uhH5nHQqASX9iioSCbMlvwLJp8NWEXS-reHWsHVv&tbo=u&sa=X&ved=0ahUKEwj7xprOjfzVAhXRIlAKHbczDooQ9C8IHA&biw=1366&bih=662&dpr=1#imgrc=Prb_LBpuEanLXM: Accessed 30 August 2017.

Web7.https://www.google.com.tr/search?q=Scan+moving+object+with+safety+camera&tbm=isch&tbs=rimg:CT62_1ywabhGpIjhZLZfUP2jwBrcyDhvajMoYWVB6ngvVYj3C9rc6aLIfczGKDL-OhT5OBY-vSKx5_1LqzJb8CyafDVioSCVktl9Q_1aPAGESB5wz6bi6RLKhIJtzIOG9qMyhgR-HI440I0jzwqEglZUHqeC9ViPRGEZYsRRNt0PyoSCcL2tzposh9zEbRRjNd3efRGKhIJMYoMv46FPk4REiUfwchoyR8qEgkFj69IrHn8uhH5nHQqASX9iioSCbMlvwLJp8NWEXS-reHWsHVv&tbo=u&sa=X&ved=0ahUKEwj7xprOjfzVAhXRIlAKHbczDooQ9C8IHA&biw=1366&bih=662&dpr=1#imgrc=tzIOG9qMyhiovM: Accessed 30 August 2017.

Web8.https://www.google.com.tr/search?q=Robot+arm+in+the+conveyor+line&rlz=1C1NDCM_trTR717TR717&source=lnms&tbm=isch&sa=X&ved=0ahUKEwih3fSSk_zVAhXDI1AKHbOnAO4Q_AUICigB&biw=1517&bih=735#imgrc=Gwd8-2K6YdVlkM: Accessed 30 August 2017.

Weiqi Li. (2011, May). "A Parallel Multi-Start Search Algorithm for Dynamic Traveling Salesman Problem." *In International Symposium on Experimental Algorithms - Springer Berlin Heidelberg*: 65-75.

West Marc. 2011. "Asset Allocation To Cover A Region Of Piracy (No. DSTO-TN-1030)," *Defence Science And Technology Organisation Edinburgh (Australia) Maritime Operations Div.*

Yan Xue-Song, Li-Shan Kang, Zhi-Hui Cai, and Hui Li. (2004, August). "An Approach To Dynamic Traveling Salesman Problem," *In Machine Learning and Cybernetics, 2004. Proceedings of 2004 International Conference on – IEEE* 4: 2418-2420.

Yousefikhoshbakht Majid, Farzad Didehvar, and Farhad Rahmati. 2016. "An Effective Rank Based Ant System Algorithm For Solving The Balanced Vehicle Routing Problem." *Int. J. Ind. Eng. Theory Appl. Pract* 23: 13-25.

Zhou Aimin, Lishan Kang, and Zhenyn Yan. (2003, December). "Solving Dynamic TSP with Evolutionary Approach in Real Time." *In Evolutionary Computation, 2003. CEC'03. The 2003 Congress on* 2: 951-957.

Zne-Jung, Leea Chou-Yuan Lee, and Shun-Feng Su. 2002. "An Immunity-Based Ant Colony Optimization Algorithm for Solving Weapon–Target Assignment Problem." *Applied Soft Computing* 2(1): 39 – 47.

In: Human-Computer Interaction
Editor: Turgut Ozseven
ISBN: 978-1-53616-495-4

Chapter 6

STATIC AND DYNAMIC MECHANICAL ENGINEERING EVALUATION OF AIS-307 STAINLESS STEEL CONSTRUCTION SCAFFOLDING-FORMWORKS SYSTEM IN FORMING GROUND CONCRETE: A COMPARATIVE EXPERIMENTAL STUDY AND ANSYS DESIGN OPTIMISATION BY SIMULATION TECHNOLOGY

Özdoğan Karaçalı*[*], *PhD
Department of Mechanical Engineering,
Istanbul University-Cerrahpaşa, Istanbul, Turkey

ABSTRACT

This paper is presenting about computer environment of formwork and scaffolding for forming the ground concrete in building construction.

[*] Corresponding Author's Email: ozdogank@istanbul.edu.tr.

When construction formwork and scaffold used in the construction of ground concrete contains many hazards and risks in terms of working conditions, the number of occupational accidents occurring and the losses caused by these accidents take the first place among all sectors. The design of the system must be realised in computer environment to see and analyse the visual of the constructed buildings before realisation of the real construction concrete by formwork and scaffolding. In the construction sector, especially in the temporary construction of scaffolding, the works have a dynamic structure that varies, the working environment can be above or below the ground level, many different jobs can be done at the same time, the sector requires combating with nature, and the variety of equipment and materials used with working methods and so on. Features distinguish the sector from other sectors. These unique characteristics of the sector lead to an increase in the number of occupational accidents and these accidents are generally severe. Therefore, due to the current situation of the formwork/scaffoldings and insecure acts and behaviors in the computer environment aided studies carried out, the scaffolding almost invites work accidents and the number of accidents falling from the scaffold increases.

This research's aim is to investigate the mechanical static analysis and cyclic load deformation properties of the under floor scaffolding system (UFS) construction components made of AIS 307 stainless steel in computer environment. UFS is a computer supported interactive analysis environment. In this paper, the method of calculation of compression and equivalent von Mises stresses has been developed in computerised 3D FEA model to develop optimsation model. Static and dynamic mechanical engineering evaluation of AIS-307 stainless steel construction scaffolding-formworks system in forming ground concrete was studied and a comparative experimental study and ansys design optimisation were presented. The results of the research work showed that the UFS material examined exceeds one million fatigue life at stress and stress levels equal to 35% of the final strength of the standard material. The experimental results confirm that the ANSYS optimisation modeling approach calculated for finite elements can be used efficiently to estimate deformation.

Keywords: scaffolding, temporary wormwork, finite element analysis, AIS 307 steel material, mechanic stress-strain, von Mises distorrion criteria, optimisation

INTRODUCTION

The under floor scaffolding system (UFS) is used as a bottom base support for the transportation of ground and beam loads. In terms of occupational health and safety, it has a critical importance in transporting loads (Buitrago et al. 2018). In this research, the general design stages of the adjustable prob system, the materials used and the forces on it, the finite element analysis (FEA) method was examined as shown in Figure 1. In assessing the fatigue life of UFS support joints; AIS 307 stainless steel material is critical in determining the number of cycles of stainless steel before fatigue. The purpose of this research is to investigate the mechanical static analysis and cyclic load deformation properties of UFS construction components made of AIS 307 stainless steel. In this, the method of calculation of compression and equivalent von Mises stresses has been developed in 3D FEA model. With the analysis of the finite element models generated in the research presented in this book, it can help design engineers and construction scaffolding manufacturers to evaluate new design strategies with UFS. This research has shown that UFS is useful for the design of construction support structures to help increase field security in understanding the current work load limits and to prevent fatal injuries and risks in the construction industry (Bernuzzi, Simoncelli, and Venezia 2017).

The method of using the temporary scaffolding system, the desired geometric structure and to support the concrete is built to support the temporary structure until it is self-supporting the support, mold operation or drawing process is the process (Michael et al. 2018). In this study, the sub-structural scaffolding system (Karaçalı 2016) is the main structural elements used to transport the floor and beam loads. Although UFS offers an ideal and economical method for all types of skeletons, slabs, columns, beams and walls, no scientific research on the problems and technical information presented in the literature has been published or is very few (Buitrago et al. 2018; Bernuzzi, Simoncelli, and Venezia 2017; Michael et al. 2018; Hemant, Prakash, and Ankush 2016; Daukšys et al. 2016; Rokade Mitul et al. 2017; Gallego et al. 2009). This makes it versatile for the use of a large number of UFSs that can be obtained between the closed and open heights. The detail of UFS fragments and research areas is shown in Figure 2. Due to their wide-working heights and high-carrying capacities, these components are under high pressure and tensile-compressive forces.

The research described in this paper addresses cyclic loading and fatigue in accordance with EN 1065 of UFS. The solution of any construction problem, such as UFS, refers to the displacement of each node and to determine the tension within each element. Analytical mathematical solutions are not suitable for solving problems related to complex geometry, loading and mechanical material properties (Rokade Mitul et al. 2017; Gallego et al. 2009; Markov and CEE 4450 2017). It is known that the elements of the UFS structure are exposed to different types of load and are taken statically during a gap load in our problem. The loading conditions can cause a stress state, such as the inability of the elements of the soil structure to function properly (Ilinoiu 2004). To make a safety assessment of the UFS structure, it is necessary to know the geometry and stress density and/or distribution due to the UFS geometry composed of the static load condition (Chaudhary 2017).

For acceptable solutions, finite element analysis is used. Numerical methods give approximate values of unknowns at a certain number of points in continuity. The modern development of FEM began in the 1940s in structural engineering. The first use of two-dimensional elements was carried out in 1956 in Turner's engineering applications. They obtained stiffness beams, beam elements and hardness matrices for two-dimensional triangular and rectangular elements. The first term occurred when the finite element was used by Clough in 1960 for the analysis of the plane stress of both triangular and rectangular elements. From the early 1950s to the present, enormous progress has been made in the implementation of the FEM to solve complex engineering problems (Alvarado et al. 2010, Karaçalı 2016).

This research aims to improve the geometry of the scaffold system structure so that it can withstand the nonlinear structural loads with maximum system mass.

Figure 1. The use of telescopic prop and formwork in building construction.

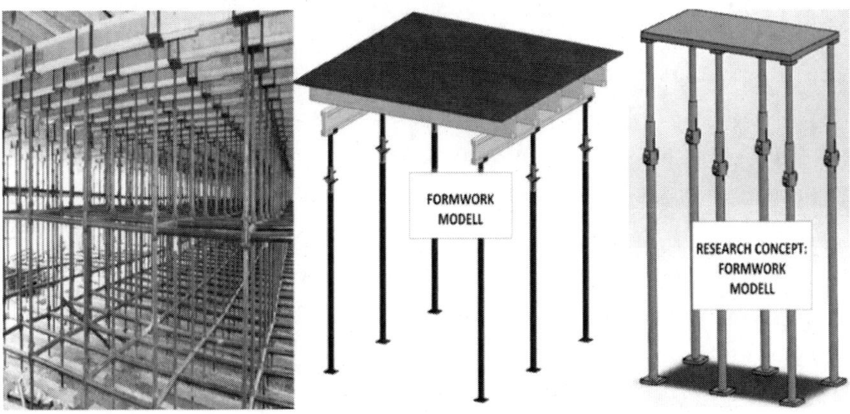

Figure 2. Pattern of formwork construction system in computer environment.

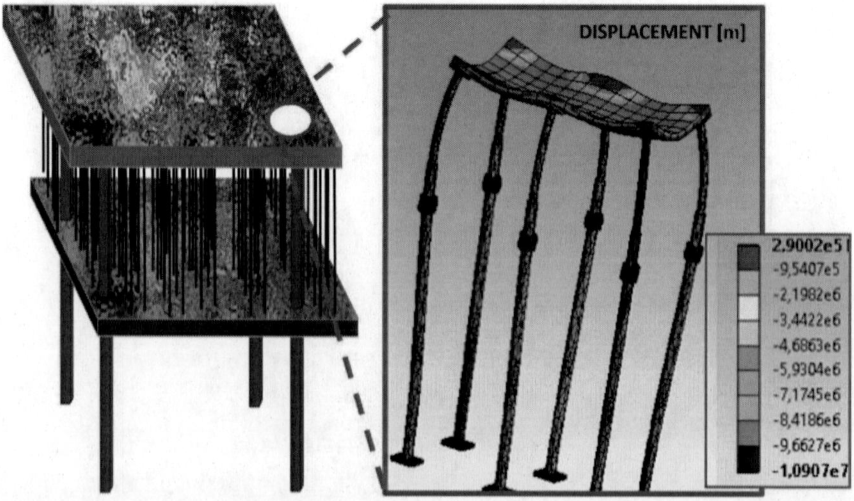

Figure 3. Computational simulation of collapse during construction in ANSYS.

A series of different geometry models for design parameters improvements were examined for formwork and scaffolding system as can be seen in Figure 2 and Figure 3. The comparison of the computational models analysis presented that the new temporary scaffold system was needed as a result of the optimal performance of these structures, the 3D modeling procedure of the entire building and the fine-tuning of geometry that aimed to form and develop the best load path. UFS is modeled on the basis of cooperation between all components as shown in Figure 3.

The construction of a concrete building often requires mold to support. These components; concrete forms, fresh concrete, construction materials, workers, equipment, various effects are sometimes designed to support loads such as wind engineering structures. The forms should always resist to all practical loads without collapse or excessive deviation, as it should be. Therefore, mold should be designed in terms of quality, safety and economy. The mold is the total and general support system for the newly prepared concrete and likewise includes all other supporting elements, hardware and covers in contact with the concrete, such as supporting. On the other hand, scaffolding form system is a temporary structure designed and built to support the works in the construction process. In addition to the steel bridge beams, there may be temporary support of precast-precast concrete elements to be tensioned together for many other applications. Mold manufacturing can be configured or customized to meet a specific design produced by an authorized person or to meet a specific system which can then be used in accordance with the manufacturer's recommendations.

Numerous mold errors are reported each year. In almost every case, a die is very uniform: failures are a result of the rapid concrete placement rate, and when the forms are used many times and the work is still underway, collapses occur during the construction phase. Several failures have been reported at the beginning of a construction project (Alvarado et al. 2010). In this study, it was supported by the scaffolding shown in Figure 3 to simplify the collapse of the staged collapse of ANSYS in order to prevent the collapse of the structure during concrete pouring. In the following section, the method of measuring stress, fatigue and damage arising from concrete loads by the finite element method developed for the UFS system is explained.

MATERIALS AND METHODS : TEMPORARY SCAFFOLDING OPTIMISATIONS SYSTEM OF CONSTRUCTION CONCRETE GROUND-UFS

The foundation of UFS, which is used in the construction of concrete floors, is easy to transport even though it takes time to dismantle. The number of repetitions is quite large, the working capacity, service life and sensitivity are high. It can be produced according to different head types for various applications. It can be adjusted with adjustment mechanism or desired height. Produced according to different working heights. In this study, the UFS finite element model was reconstructed from a normal industrial data set. The model was rebuilt and rigorously examined for mechanical responses to physical ground loads. Since the entire UFS region has a specific interest, each part is taken into account in the analysis. The main research objectives are to create a model to improve site safety to understand the current working load limit for the design of UFS structures and to avoid the risk of fatal injury in the construction industry. In order to compensate for stress and stress distribution in the material, a real laboratory test environment and simulation model including static and dynamic analyzes was created. The test is carried out according to DIN EN ISO/IEC 17025: 2005 standards. Joint testing configurations were conducted to investigate the strength and strength of the UFS connections in the vertical direction (rotating around the z-axis) and in the horizontal direction (rotation around the y-axis). Loading directions occur up or down to be bent around the y-axis. Loading directions consist of up or down to bend around the y-axis. By mechanical testing of the load overflow capacity, deformation of the metal joints, scaffolding, formwork and shelf systems working with the scaffolding systems and scaffolding structure service elements of UFS were observed. The UFS was coarse adjusted with the fine tuning means using a pin and a threaded sleeve inserted into the holes in the inner tube. The telescopic stand has a rotating handle that facilitates rotation in limited areas.

Static ground load is considered to be the most critical load that can lead to fatigue problems of UFS. (Daukšys et al. 2016). In the same way, the dead (the weight of the concrete to be poured) is a dynamic load with a large number of life cycles, as the weight is a permanent load and contributes significantly to the stress distribution. Static analysis is based on finite element analysis (FEA) models and is intended to calculate deformation and stresses under operational conditions such as static load. The deformations of the KIA components were used to define the required maximum range of the positions of the UFS and the static misalignments had to be corrected by simulation. In addition, the rigid body displacements and rotations of the UFS components were calculated based on the von Mises equivalent stresses. This has been very useful for further design improvements and optimization work. In order to fully estimate the mechanical load behavior of the UFS material, it was necessary to understand the physical phenomena to determine the general properties by means of SEM. As a result, the development of quantitative displacement and stress distribution was presented by the deformation model method under static loads as shown in Figure 4 and Figure 5.

Very high stresses have occurred in the frame structure under the gravity load on the PDP. The maximum Von Mises occurred in an area under the UFS's supporting structure, highlighting the order of 680 MPa. In order to meet the mechanical stresses in UFS, some measures have been taken in this research: (1) The frame structure of critical areas has been strengthened by increasing the cross-sectional areas of critical areas, (2) The topology of the UFS framework structure has been changed. As shown in Figure 6, the yield strength limits of the steel materials used in critical regions (3) were found to be 427 MPa for AIS 307.

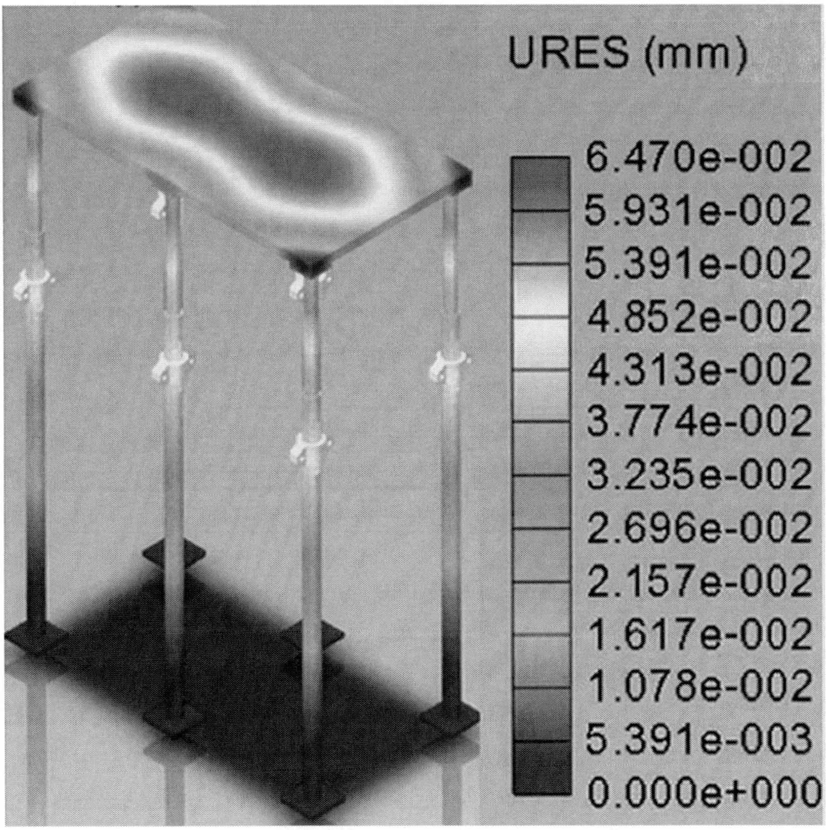

Figure 4. Deformation model under static load scaffold system static loads.

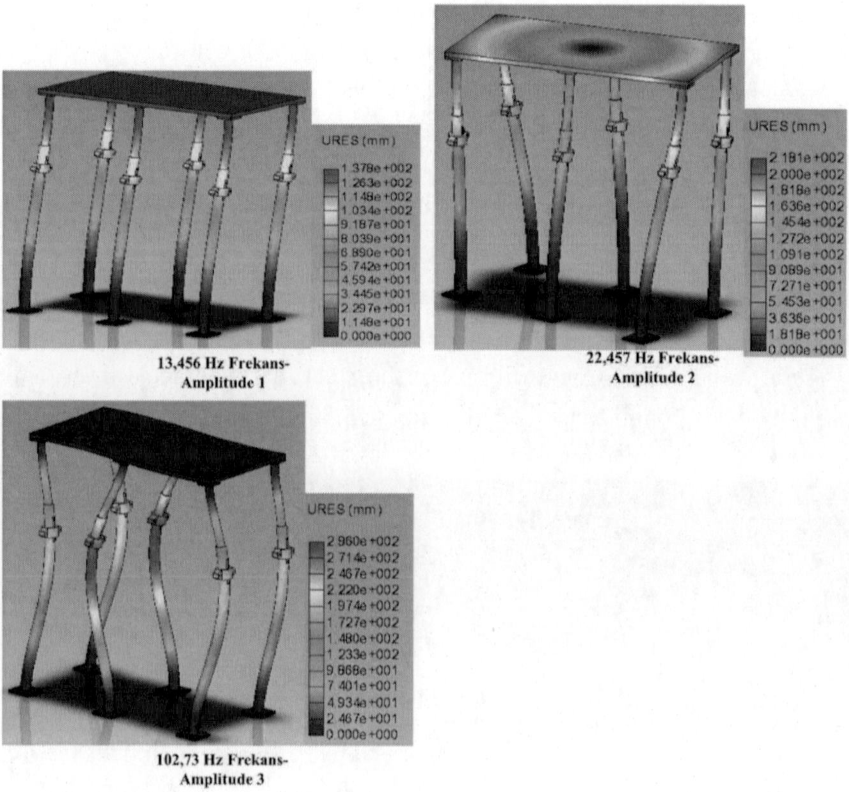

Figure 5. Deformation analysis: in the sub-skeletal system at different frequency and amplitude in computer environment.

The 3D finite element models developed using the ANSYS software formed the half of the tube and the connecting assembly as shown in Figure 5, since the UFS had a symmetry plane along a section along its longitudinal axis. Material, geometry and energy-based convergence criteria - von Mises method was used. The results from UFS for load deflection zones were associated with the outcomes of the experimentation; this displayed that UFS narrowly forecasted UFS performance with a highest error of 1.4% for AIS 307 steel. The reliability factor of the FEA model, the end plate of UFS, the length, the cross-section, the bolt head, the bolt handle and the size and shapes of the bolt nut were analyzed as in Figure 7 from the exact values of the UFS geometric parameter values. The key factor, the modulus of elasticity E, was reevaluated as shown in Figure 8 for yields and equivalent von Mises stresses.

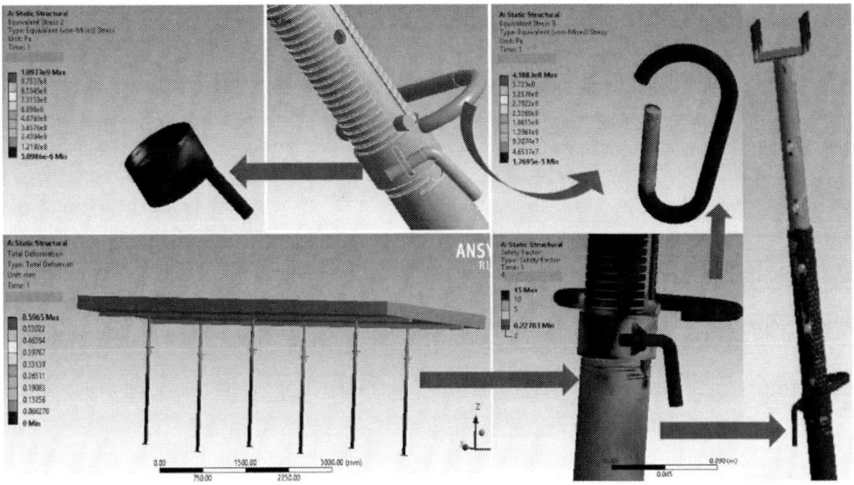

Figure 6. The von Mises analysis of the telescopic prop system in the construction formwork.

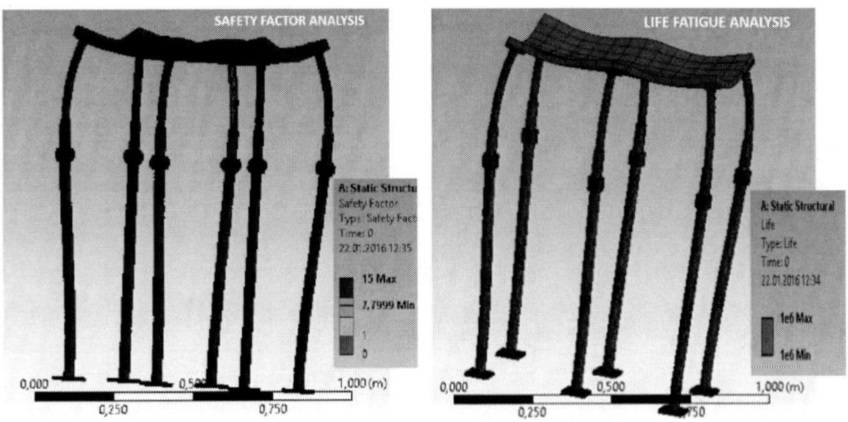

Figure 7. Analysis of the safety factor of the molded skeleton system.

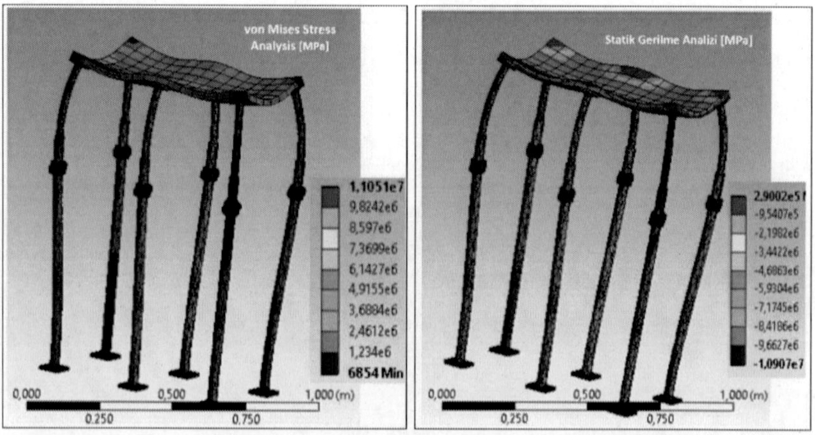

Figure 8. von Mises and static stress analysis of base in formwork system.

In this study, a very useful fatigue analysis has been applied for cyclic loading, providing the required fatigue life to be reached. The deviation areas corresponding to the stresses due to the loading in UFS provide damage-tolerant design. The area of mechanical stress, which leads to fatigue life, which is permitted and not allowed, has been investigated.

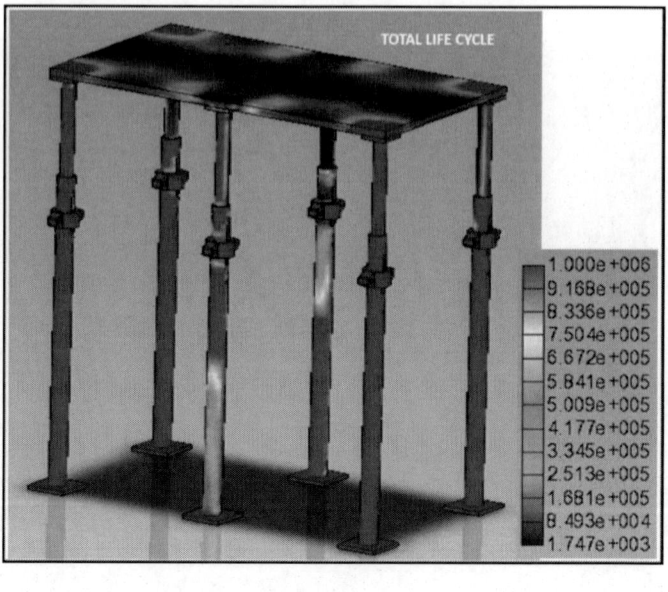

Figure 9. Life-cycle analysis of the formwork system.

The basic scope of the dynamic analyzes allowed us to estimate the dynamic behavior of the PDI structure. AIS 307 of the material of UFS has analyzed the irregularity of mechanical stresses and calculated geometric shapes. In UFS, laboratory tests were carried out representing the distribution of hazards to control fatigue resistance when cyclically loaded at the specified maximum strength amounts. UFS fatigue test requires material stress information that can be obtained from the work hazards observed in practice (Alvarado et al. 2010).

This has enabled the ability to prevent stresses and UFS fatigue tests under specified cyclic loading conditions. (1) The evaluation of the connection between the fatigue life (2) and the stress levels based on the concrete load affected, helped to establish a correct model using these parameters to provide a realistic basis for the design of CAI. Static fatigue tests are arranged on a universal testing machine according to ASTM standards. The work involved repetitive bending, compression, and extension throughout the UFS. The investigation of the load test of UFS and components includes hydraulic pressurization and bending tests up to the loads of several tons. What is critical for the success of a dynamically loaded structure is the fatigue design, which is based on an understanding of the fatigue behavior at all $R = \sigma min/\sigma max$ ratios. Average tensile stress was obtained for multiple different R ratios. In Figure 9, ANSYS analysis results were obtained in fatigue test.

COMPUTER AIDED FINITE ELEMENT ANALYSIS OF UFS

Most engineering problems in elasticity theory are explained by partial differential equations with appropriate boundary conditions and/or variation methods. Differential equations that define the complex geometry shapes and complexity of stress and strain for complex boundaries often cannot be solved analytically, so approximate solutions for such problems can be found by applying numerical methods (Karaçalı 2016). FEM analysis is one of the most used methods in solving many engineering problems in civil engineering. The fatigue analysis shown in Figure 8 for the modeling of the undercut skeleton system in the continuity of the plane tension and strain conditions was carried out using finite elements in this study.

In the plane of deformation, the material is uniformly deformed in all planes parallel to the x, y plane. The internal distribution of displacement in the finite element itself is defined by the interpolation functions. By using the SEM analysis method, it allowed approximation of the displacement area. The mesh nodes of x-y plane are formed which have displacements with the components. A comparison between the calculated and permissible stresses, ANSYS fatigue analysis shown in Figure 8 for the fatigue behavior in the UFS test was calculated to show the mean and amplitude stress for each node in the structure based on the tensile strength and the vibration force allowed for the strength limits of the materials. Red and green dots are upper and lower stress values.

This review addresses the design requirements that take into account the effects on the bending forces that affect the stability and durability of UFS. The fatigue problem of UFS is the crack or deterioration of the component material due to the grout fill and the cyclic stresses applied under the final strength of the material during laying. Fatigue behavior, UFS failure mode, which shows the critical stresses von Mises continuous strength of the stresses that take into account the three-dimensional isoparametric finite element model UFS model, geometric contact effects between the non-linear matching behavior is developed as shown in Figure 8. The 9845 nodes of elastoplastic solid elements were developed for 3D finite element model.

RESULTS AND DISCUSSIONS

UFS has presented the stresses resulting from continuous loads as a means to predict fatigue and damage that may occur. Stress level where the desired fatigue life was not reached within the predicted re-damage level was not reached. In addition, the geometry of the formwork with material linearity of every part of components of the UFS system remained within the limits.

This research can help increase site safety to understand the current work load limit in the design of the structures of the UFSs and makes this design consistent to prevent fatal injuries in the risky construction sector. In the clamping of the beams placed on the leg of the UFS, serious incidents can occur that form the floor support. The results of the systematic study are useful for cyclic load modeling of the UFS systems, and the results of this research can support the UFS designers and manufacturers to provide better and safer upholstery support products.

CONCLUSION

In this study, an experimental test and simulation study (Figure 8) to determine optimisation of formwork and scaffolding, fatigue behavior of UFS, mechanical and static analysis method has been proposed and examined. This article presents a new method of FEA cyclic loading behavior related to fatigue analysis using experimental and ANSYS computational engineering software. The UFS steel sample sent to the tensile, compression and torsional strength test was applied to the UFS steel sample, based on the cyclic loading and compression forces finite element methodology. Finite element analysis - ANYS-based research and experimental study, AIS 307 was made on steel materials and adjustable steel materials that can be adjusted to determine static steel loads and load deviation behavior under static and dynamic loading and boundary conditions set from the specified model. This optimisation approach in the study revealed that the maximum deviation occurs in the upper part of the CAE and that the maximum stresses are concentrated at the intersection with the telescopic load. The largest displacement of UFS concrete support system was 0.07 m. 0.025 m displacement amount was measured under high pressure after wind and concrete loading. With the finite elements made in ANSYS, the calculation was 0.26 m. This approximate calculation proved the accuracy of the study and the UFS system to be safe. As expected, the most important parameter in terms of critical loads was wind, wind direction 300, and rain. In this study, only the wind effect was evaluated.

The results of the study showed that the UFS material examined exceeds one million fatigue life at stress and stress levels equal to 35% of the final strength of the standard material. The experimental results confirm that the ANSYS modeling approach calculated for finite elements can be used efficiently to estimate deformation.

ACKNOWLEDGMENT

This research of 2019-58846 project was supported by Istanbul University-Cerrahpaşa BAP-scientific research unit.

REFERENCES

Alvarado, Yezid A., Pedro, A. Calderón., Isabel, Gasch. & Jose, M. Adam. (2010). 'A Numerical Study into the Evolution of Loads on Shores and Slabs during Construction of Multistorey Buildings. Comparison of Partial Striking with Other Techniques'. *Engineering Structures*, *32* (10), 3093–3102.

Bernuzzi, Claudio., Marco, Simoncelli. & Marcello, Venezia. (2017). 'Performance of Mono-Symmetric Upright Pallet Racks under Slab Deflections'. *Journal of Constructional Steel Research*, *128*, 672–686.

Buitrago, Manuel., Juan, J. Moragues., Pedro, A. Calderón. & Jose, M. Adam. (2018). 'Structural Failures in Cast-in-Place Reinforced Concrete Building Structures under Construction'. In *Handbook of Materials Failure Analysis*, 153–170. Elsevier.

Chaudhary, Aakanksha Dilip. (2017). 'Study of Tunnel Formwork System & Comparative Analysis with Conventional Formwork'. *International Journal of Scientific & Engineering Research*, *8* (5), 2229–5518.

Daukšys, Mindaugas., Gintas, Brazas., Jolanta, Šadauskienė., Mindaugas, Augonis. & Šarūnas, Kelpša. (2016). 'Productivity Analysis of Concrete Slab Construction by Using Different Types of Formwork'. *Journal of Sustainable Architecture and Civil Engineering*, *15* (2), 38–46.

Gallego, E., Fuentes, J. M., Ramirez, A. & Ayuga, F. (2009). 'Design of Tall Formworks by a Finite-Element Model'. *Journal of Construction Engineering and Management*, *136* (7), 803–810.

Hemant, Rawal Meet., Bharakhada, Mayank Prakash. & Dhawde, Akash Ankush. (2016). 'Shuttering Design & Cost Comparisons'. *International Journal of Recent Advances in Engineering & Technology*, *4* (3), 2347–2812.

Ilinoiu, George. (2004). 'Wall Formwork Design'. *Civil Engineering Dimension*, *6* (2), 101–108.

Karaçalı, Ö. (2016). 'A New Perspective on Cyclic Loading Behavior Analysis of ATSP-Adjustable Telescopic Steel Prop S235GT Material Used in Structural Engineering', *International Journal of Acta Physica Polonica A*, vol. *129*, no. 4, pp. 429-436.

Markov, Ivan. & Students of CEE 4450. (2017). 'Educational Modeling and Load Testing of Shells'. *International Journal of Space Structures*, *32* (1), 53–65.

Michael, Jacob., Scott, H. Smith., Stephan, A. Durham. & Mi, Geum Chorzepa. (2018). 'Crack Control in Concrete Walls through Novel Mixture Design, Full-Scale Testing, and Finite Element Analysis'. *Construction and Building Materials*, *166*, 301–314.

Rokade Mitul, R., Bhor Nikhil, S., Virkar Aniket, K., Rode Aksahy, R. & Maid Nilesh, S. (2017). 'A Paper on Design of Formwork'. *International Journal of Engineering Sciences & Management Rokade*, *7* (1), 2277 – 5528.

In: Human-Computer Interaction
Editor: Turgut Ozseven

ISBN: 978-1-53616-495-4

Chapter 7

Geological Strength Index (GSI) Determination by Local Image Descriptors and Machine Learning Methods

Ahmet Selman Bozkir[1,*], Hakan Ahmet Nefeslioglu[2], Ogan Kartal[3], Ebru Akcapinar Sezer[1] and Candan Gokceoglu[2]

[1]Department of Computer Engineering, Hacettepe University, Ankara, Turkey
[2]Department of Geological Engineering, Hacettepe University, Ankara, Turkey
[3]General Directorate of Turkish State Railways, Ankara, Turkey

Abstract

The main purpose of this book chapter is to develop a computer vision based approach which is able to calculate approximate Geological Strength Index (GSI) value of rock mass when the photograph of rock outcrop is given to the computer. For this purpose, local image

* Corresponding Author's Email: selman@cs.hacettepe.edu.tr.

descriptors Scale Invariant Feature Transform (SIFT) and DAISY and machine learning methods Support Vector Regression (SVR), Random Forests (RF), and XGBoost were investigated. The chapter involves three mains sections as follows: (i) The details of GSI are described in the "Geological Strength Index" section; (ii) Local image descriptors and machine learning methods investigated in this study are summarized in the "Methods" section; and (iii) the details of data, descriptor generation, prediction models, and validations were provided in the "Experimental Studies" section. The GSI values defined for eleven rock types at twenty outcrops were evaluated in the experimental studies. The total number of digital images collected for our corpus is 1488. As a result, by using the SIFT descriptors and XGBoost estimator, the accuracy values were obtained as higher than 99% and 90% for training and test stages, respectively. This promising result shows that the proposed method is quite acceptable and computer vision methods are capable to estimate GSI value of a rock mass from photo.

Keywords: Geological Strength Index, Local Image Descriptors, SIFT, DAISY, Machine Learning

INTRODUCTION

In terms of engineering geology, rock mass involves two main components such as intact rock and rock mass discontinuity. According to the definition given by Zhang (2005), intact rock can be defined as a continuum or polycrystalline solid between discontinuities consisting of an aggregate of minerals or grains, while rock mass is the environment composed of intact rock materials separated by discontinuities. On the other hand, the term discontinuity is defined as any separation in a rock mass having zero or low tensile strength (Zhang, 2005). Evaluation of rock mass environment is the fundamental issue in rock engineering applications. Assessment of a rock mass environment involves the evaluations of strength and deformation characteristics of whole rock environment including intact rock and rock mass discontinuities together. However, the problem here is the specimen size required for necessary laboratory tests. Unfortunately, there is not yet a test equipment to test a sample having rock mass size in the laboratory. This constitutes the most important problem in the evaluation of the strength and deformation properties of rock masses. In order to compensate this situation, Hoek and Brown (1980) proposed an empirical failure criterion. The strength of rock mass is able to be evaluated by using this criterion. The criterion essentially based on the quality of rock mass, the properties of rock mass discontinuities in particular, and intact rock material properties such as uniaxial compressive strength (σ_{ci}) and material constant (m_i). In the early stage, in order to evaluate the quality of rock mass in the criterion, the rock mass rating (RMR) procedure was implemented (Hoek and Brown, 1988). However, the RMR procedure is not representative for the evaluation of rock mass quality, particularly in case of RMR < 30 which defines poor quality rock mass conditions. By highlighting this problem, Hoek and Brown (1997) suggested a new scheme namely Geological Strength Index (GSI) in order to evaluate the quality of rock mass more reasonable. The blockiness state and rock mass discontinuity conditions constitute the main components of the GSI scheme. In order to determine the GSI value of rock mass, the judgements of an experienced engineering geologist are essential. In the GSI scheme suggested by Hoek and Brown (1997), rock mass was evaluated by considering twenty classes based on visual interpretations of blockiness and rock mass discontinuity conditions. The GSI values varied between 10 and 80 on the first scheme. Later, additional classes for foliate/laminated/sheared and massive or intact rock environments were also integrated into the GSI scheme (Hoek et al., 1998; Hoek, 1999). The creators of the GSI scheme stated that the determination of GSI values should not be so precise; instead of this, considering inherent uncertainty of rock environment, GSI values can be defined by quoting a range (Hoek and Brown, 1997; Hoek et al., 1998; Hoek, 1998; Hoek 1999). There are more recent GSI schemes, details of which are given in the further sections of this chapter. On the other hand, some investigations were also carried out in order to determine GSI values more precisely (Sonmez and Ulusay, 1999; Cai et al., 2004; Russo, 2007; Russo, 2009; Hoek et al., 2013). Marinos et al., (2005) expressed the main reason for these attempts as a sort satisfaction of the perception that engineers are happier with numbers. It is obvious that considering inherent uncertainties of rock environment, precise determination of GSI values may not be reliable. The main reason for this peculiarity is the restrictions on the evaluations of the parameters included in these quantified schemes; the blockiness evaluations and assessments of discontinuity conditions may not be as precise as they are expected. Considering these issues, in order to compensate uncertainties and expert subjectivities, a new GSI determination method is introduced in this book chapter. For this purpose, the digital images of rock outcrops of which GSI values were determined by experienced engineering geologists were taken into consideration.

In this study, visual features of the digital images belonging to rock outcrops were extracted by using local image descriptors namely Scale Invariant Feature Transform (SIFT) (Lowe, 2004) and DAISY (Tola et al., 2010). The revealed features were then quantized into feature vectors by bag-of-visual-words approach and then were fed into machine learning methods (i.e., SVR, Random Forest, XGBoost) in order to predict GSI values of the rock outcrops. The prediction performance of each descriptor and machine learning method pair have been then compared and discussed.

This chapter includes three mains sections as follows: (i) The details of GSI are given in the "Geological Strength Index" section; (ii) Local image descriptors and machine learning algorithms implemented in the present study are described in the "Methods" section; and (iii) the details of data, descriptor generation, prediction models, and validations were given in the "Experimental Studies" section.

Geological Strength Index

Why is the GSI scheme so important? First of all, let us explain this issue in more detail. Rock mass classifications e.g., the RMR system suggested by Bieniawski (1973) and the Q system proposed by Barton (1974) are not reliable under complex ground conditions such as squeezing, swelling, spalling, slabbing, and rock burst under high stress levels. These systems are essentially suitable for the situations in which the rock mass behavior is simple that RMR values vary between 30 and 70 under moderate stress levels (Marinos et al., 2005). It is obvious that numerical methods can be a solution for this problem and used for the analysis of these conditions. However, the main requirement of the numerical methods is reliable inputs defining strength and deformation characteristics of rock mass. As mentioned previously, considering necessary specimen size, it is almost impossible to determine strength and deformation characteristics of rock mass in the laboratory. This situation is the inevitable reason for the development of the Hoek-Brown rock mass failure criterion; thereby the rock mass failure criterion was suggested by Hoek and Brown (1980). In the earlier version of the criterion, there were difficulties in determining the rock mass constants (m and s). For this reason, the creators of the criterion mentioned that geological observations which was able to be performed quickly and easily should be the base of the evaluations. By considering this situation, Hoek and Brown (1988) integrated the RMR classification system to the criterion. However, as stated above, not only the RMR classification system but also the Q system both suffer from the applications performed in poor quality rock masses. The main reason for this inadequacy is the Rock Quality Designation (RQD) parameter which was first introduced by Deere (1964) as a separate classification system. The systems RMR and Q both involve the parameter. However, the RQD values are mostly zero for poor quality rock masses or have no meaning (Marinos et al., 2005). This is why the GSI scheme is so important. The scheme was introduced in order to compensate the uncertainties and restrictions described up to now.

The recent version of the GSI scheme (Hoek et al., 1998) is given in Figure 1; and the GSI scheme for heterogeneous rock masses suggested by Marinos and Hoek (2001) is given in Figure 2. The main component of the GSI scheme is in point fact engineering geologist. The engineering geological evaluations performed for rock masses with respect the scheme are essentially qualitative. As discussed previously, precise determinations may be meaningless when considering complex rock environments. In the method proposed in this study, we are actually trying to make.

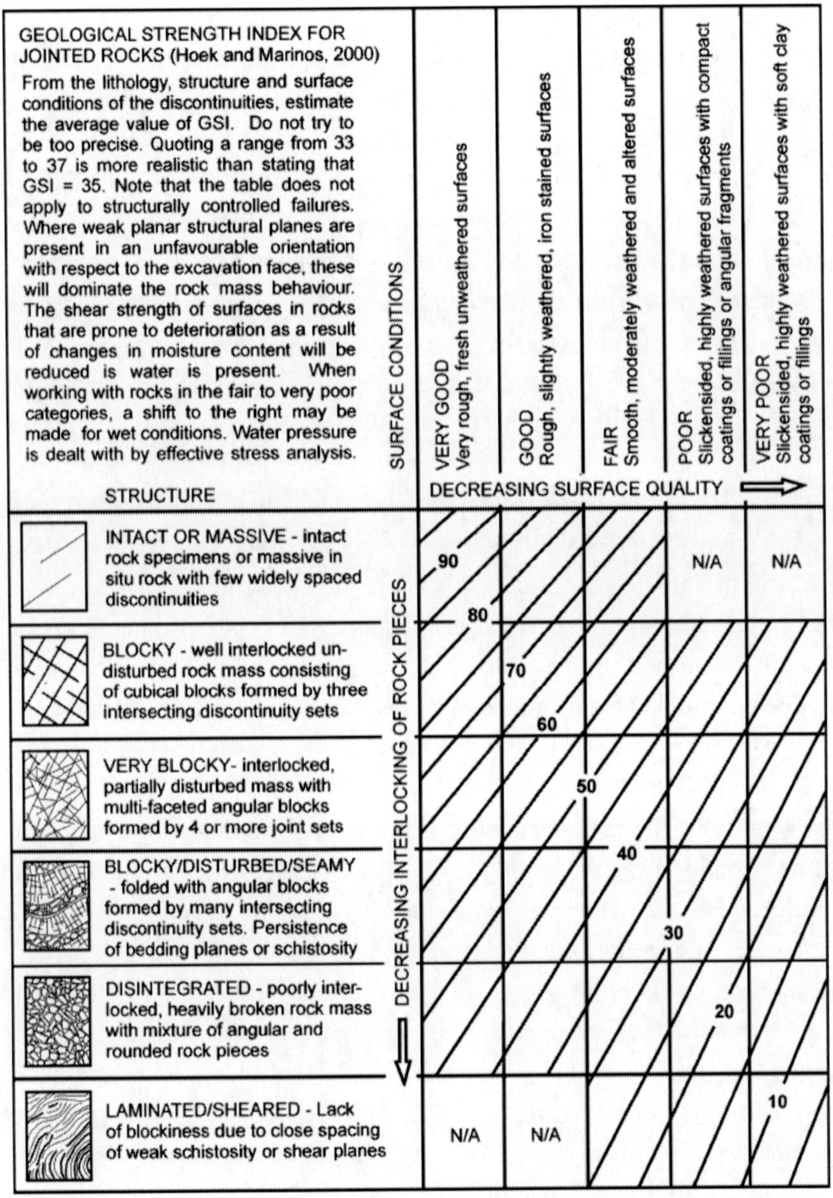

Figure 1. The general GSI scheme proposed by Hoek et al., (1998).

The computer think like an engineering geologist. When determining the GSI values according to the schemes given in Figure 1 and Figure 2, an engineering geologist intuitively evaluates the blockiness state and the discontinuity conditions of rock mass based on his/her experience. These evaluations are based on visual examinations of the rock mass exposed in outcrops. In the proposed method, the digital images of rock environments correspond to the sight observed in field by the experienced engineering geologist. The image descriptors generated from the digital images denote the features extracted intuitively from rock environments by experienced engineering geologists.

And lastly, the machine learning and estimation processes performed by using image descriptors correspond to the determination of GSI values by the experienced engineering geologist. It should be paid attention here that the proposed methodology is not a sort of quantification. Rather than quantification, the method suggested in this study is automatization of GSI determinations.

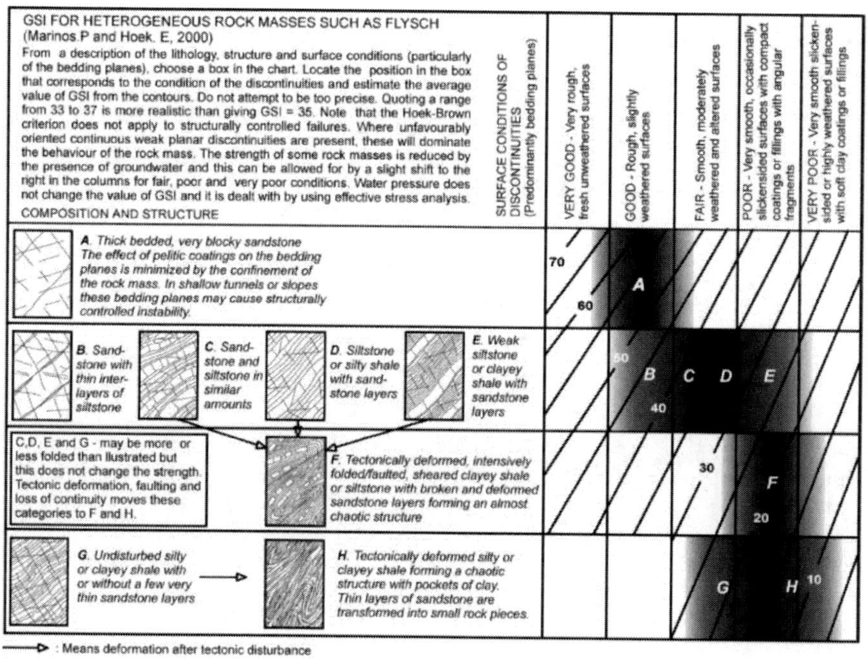

Figure 2. The GSI scheme for heterogeneous rock masses suggested by Marinos and Hoek (2001).

METHODS

Local Image Descriptors and Representation Methodology

In this section, we provide a background for local image features and machine learning methods we have employed during this research study. In order to verify whether GSI estimation is prone to rotational invariance, we have used two different image descriptors one having rotational invariance (SIFT) property whereas the other (DAISY) has not.

Scale Invariant Feature Transform (SIFT)

Invented by Lowe (2004), Scale Invariant Feature Transform (SIFT) aims to generate highly distinctive key point based descriptors which scale, rotational and translational invariant. Moreover, as stated by Sachdeva et al., (2017) SIFT descriptors are robust to geometric transformations and present satisfactory results in task of near-duplicate image retrieval (Foo and Sinha, 2007). Since its discovery, mostly related to computer vision related applications, SIFT has found a wide variety of usage in fields such as image identification (Karami et al., 2015), vehicle logo recognition (Keser et al., 2017), wireless sensor node localization (Fuiorea et al., 2008) and phishing web page detection (Afroz and Greenstadt, 2011)

In simple terms, the SIFT algorithm extracts 128D feature vectors describing the local context at keypoints found by any key point detection method (e.g., Harris corner detector). Similarly, according to Karami et al., (2015), during the SIFT computation, an input image is being transformed into a collection of local feature vectors. Since the descriptors are generated around key points, SIFT features are being considered as local image descriptors.

Generation of SIFT descriptors involves four stages: (1) scale space detection, (2) key point localization, (3) orientation assignment and (4) keypoint description. In order to obtain scale space filtering throughout the visual perception, Laplacian of Gaussian (LoG) can be used with variour sigma values. It is noteworhty that, by defining different scaling parameter of sigma, LoG can be used as a blob detector to detect varying sized blobs. That being said, Gaussian kernel with low sigma values generates high score for small corners whereas high sigma values suit better for larger corners. In this way, local maxima points among different scales of the image can be achieved. Nonetheless, in design of SIFT algorithm, Difference of Gaussians (DoG) has been selected instead of LoG since it is a low cost approximation of LoG in order to find list of (*x,y, sigma*) tuples denoting potential key points at respective sigma values. Throughout the DoG computation, the input image is deliberately blurred with two different successive sigma values yielding different octaves. By taking the difference of newly generated blurred images a pyramid over scale space is formed. Next, each pixel in middle layers of pyramid is compared against 26 (i.e., 9 + 8 + 9) neighbor pixels including upper and lower layers. Thus, local extrema values are marked as possible key points (Keser et al., 2017). The paper of Lowe (2004) state some empirical results for defining optimum octave and scale levels such as 4 and 5 respectively. In this configuration the sigma has been detected as 1.6. The formulas for calculating the gaussian pyramid is given in expressions below:

$$L(x, y, \sigma) = G(x, y, \sigma) * I(x, y) \tag{1}$$

$$G(x, y, \sigma) = \frac{1}{2\pi\sigma^2} e^{-(x^2+y^2)/2} \tag{2}$$

$$D(x, y, \sigma) = (G(x, y, \sigma) - G(x, y, \sigma)) * I(x, y) = L(x, y, k\sigma) - L(x, y, \sigma) \tag{3}$$

As pointed out by Karami et al., (2015), scale-space extrema detection procedures generate so many candidate keypoints potentially causing unstable results and need to be filtered out. Moreover, DoG is more sensitive to edges. Therefore, in order to do a refinement over the detected key points and obtain more useful key points, the key point localization stage is carried out. In this step, a quadratic Taylor expansion of the DoG along with candidate key points is first computed and an interpolation is made to accurately detect more useful keypoints and suppress edge based key points. During this computation, the Taylor expansion given below is used where *D* and its derivatives are calculated for each candidate keypoint (Lowe, 2004).

$$D(x) = D + \frac{\vartheta D^T}{\vartheta x}\boldsymbol{x} + \frac{1}{2}x^T\frac{\vartheta^2 D}{\vartheta x^2}\boldsymbol{x} \tag{4}$$

Next, based on local image gradient directions, the stage of orientation assignment is carried out. This step is crucial for achieving rotational invariance for the generated descriptors. In other words, according to Lowe (2004), "by assigning a consistent orientation to each key point based on local image properties, the key point descriptor can be represented relative to this orientation and therefore achieve invariance to image rotation". At this point, given an image instance *L(x,y)* at scale σ and gradient magnitude m(x,y) the orientation *θ(x,y)* is calculated by (5) and (6).

$$m(x,y) = \sqrt{(L(x+1,y) - L(x-1,y))^2 + (L(x,y+1) - L(x,y-1)^2} \tag{5}$$

$$\theta(x,y) = tan^{-1}\left(\frac{L(x,y+1)-L(x,y-1)}{L(x+1,y)-L(x-1,y)}\right) \tag{6}$$

As a result, for each keypoint, by using gradient directions of sample points around the key point, orientation histogram is generated. Original design of SIFT algorithm involves 36 bins to cover 360 degree range of orientations (Lowe, 2004). Following this process, every instance which have been added to the histogram is weighted by the gradient magnitude it has. For this weighting schema, a gaussian weighted circular window having a σ which is 1.5×scale of the key point is used. It should be noted that, for each examined area, the SIFT performs sampling in a 16×16 pixel wide area of a regular grid in order to cover the local region (Sachdeva et al., 2017). Further, the gradient orientations are computed from 4 × 4 patches of gradient histograms having 8 bins each that is yielding a 128 dimensional key point descriptor. The resultant descriptor is robust against scaling, rotation and translation. The figure 3 below illustrates how the SIFT algorithm finds important keypoints on our test image.

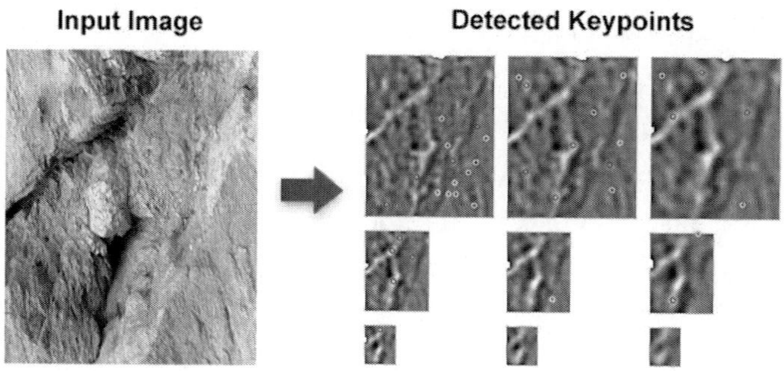

Figure 3. SIFT key point generation on a test image by using various scale spaces. The red and yellow circles show extremas. However, the yellow circles will be discarded since SIFT considers them as noisy points.

DAISY Descriptor

Invented by Tola, Lepetit and Fua (2010), the DAISY descriptor - beside inspired from SIFT - is a local image descriptor that aims fast dense feature extraction which is useful for bag-of-visual-words or similar representations. Moreover, DAISY is not only robust to photometric transforms like brightness and contrast variations but also robust to occlusions and perspective effects that view-point alterations cause (Tola, 2010). According to Tola, Lepetit and Fua (2010) it is noteworthy that DAISY is the first algorithm that primarily attempts to estimate dense depth maps from wide-baseline image pairs. In contrast to SIFT, DAISY does not involve any interest point detection thus it focuses on only dense feature extraction for various purposes such as depth map estimation from stereo images (Tola, Lepetit and Fua, 2010), visual object detection (Zhu, Bichot and Chen, 2011), discovering dense correspondences between two images (Yang, Lin and Lu, 2014) and detection of image duplication forgeries (Guo, Liu and Wu, 2013).

As can be inferred from the Figure 4 presented below, the name of the descriptor refers to the flower *daisy* shape like region of interest for a pixel. Fundamentally, DAISY is equipped with 6 configuration parametes: (a) R (*radius*) – distance from the center of the inner most circle to the center of the outer most outer circle, (b) Q (*radius quantization no*) – number of convolved orientations, (c) T (*angular quantization no*) – number of histograms for a single layer, (d) H (*histogram quantization no*) – number of bins in the histogram, (e) S (*grid point no*) number of histograms employed for generation of the descriptor = Q×T +1 and finally (f) D_s (*descriptor size*) – the length of the descriptor vector = S×H (Tola, Lepetit and Fua, 2010). The resultant descriptor size is dependent on Q, T and H values. The radius for each circle is proportional to the standard deviation belonging to the Gaussian kernels. Besides, the '+' sign shows the places where the final descriptor is generated from the convolved orientation maps centered at. Observed overlapping regions enable DAISY to have rotational invariance to an extent. Furthermore, the situation of increasing radii of the outer rings has been reported as a necessity to have an equal sampling of the rotational axis (Tola, 2010).

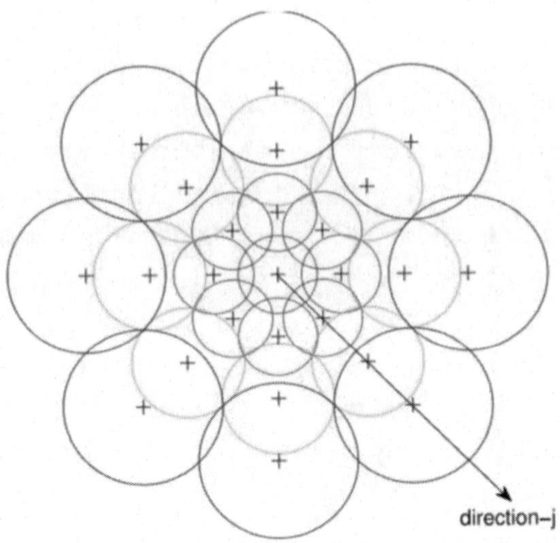

Figure 4. The central-symmetrical circle structure of ROIs in DAISY. Adopted from (Tola, 2010).

The main idea behind the computation of DAISY is to convolve the different directional diagram of original image with Gaussian filtering function with different size (Li, 2014). Actually, around the red center point, an enlargement of 3 layers having different radius is created, where 8 sampling points exist for each layer. These points have been marked with '+' symbols and distributed having distribution of 45 degree intervals. As reported by Li and Zic (2014), the sampling points per layer involve the same Gauss value scale yielding the gauss–scale–value progressively grows in a center to outside fashion. This structure and design make the DAISY descriptor to present better robustness for affine and illumination variations on images. On the other hand, though the DAISY approach involves aforementioned invariances, it does not literally provide a powerful rotational invarance. Apart from SIFT which apply rectangular neighborhood, DAISY descriptor benefits from its circular neighborhood since it represents better localization abilities.

Due to the limited number of pages we have and the amount of details regarding the generation of DAISY descriptors, we have left those details for further investigation of readers.

Bag of Visual Words Represention

Bag of visual words (BOVW) is a visual feature based representation technique that has been used in image classification problems. Note that, the BOVW concept is originated from the well known bag of words (BOW) schema which has been employed in the fields such as information retrieval, text classification and natural language processing. In principle, computation of BOW involves following stages: (a) counting the number of each word/term appears in a document, (b) calculation of frequency for each word that will be included in vocabulary and (c) generating a frequency histogram from it. In simple terms, by using BOW approach a document is transformed to a histogram based representation. Similarly the same concept also exists in BOVW representation, but instead of words, extracted image features are accepted as the "words."

The general aim of BOVW is to represent an image as a set of visual features. In most of the cases, features involve key points and descriptors. It should be noted that, regardless whether it is extracted around a key point, any kind of local image descriptor can be used in this stage. In practice, various numbers of descriptors are extracted from images. Next, we apply the stage called as quantization referring to the clustering based (e.g., K-means) visual vocabulary generation is carried out. Based on the discovered clusters, each descriptor from the image is assigned to a cluster centroid in order to build a frequency histogram of features. In this way, the created frequency histogram can be used for specific tasks such as image classification, feature extraction and information retrieval. The above mentioned steps were illustrated in the Figure 5 presented below.

Supervised Machine Learning Methods Employed

In this section we have briefly introduced the machine learning methods and the evaluation metrics we have used throughout the study.

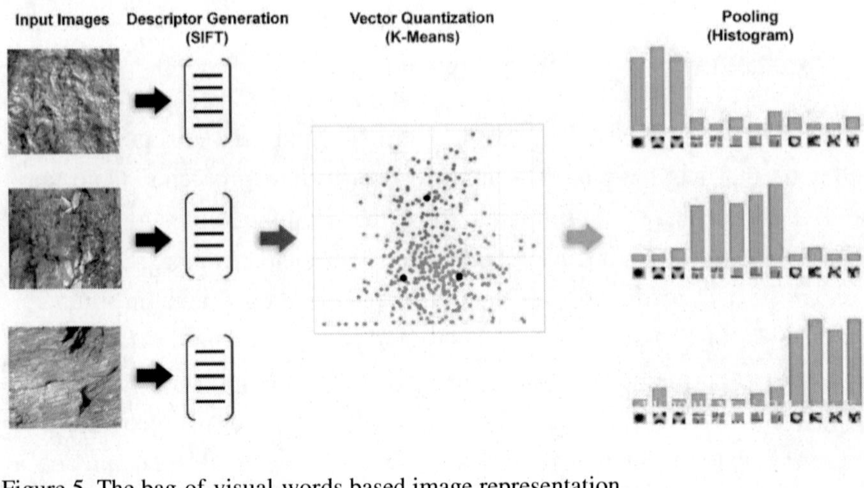

Figure 5. The bag-of-visual-words based image representation.

Support Vector Regression

Support Vector Regression (SVR) is an algorithm that is derived from well known support vector machine (SVM) classification scheme first introduced by Vapnik (1992) in order to make prediction on continuous values rather than "hard" class labels. Being a nonparametric method, SVR is a suitable regressor that can be used for linear and nonlinear seperable datasets. In principle, SVR utilizes the same principles as the SVM for classification, with only a few minor differences. Furthermore, support vector regression shares similarity with linear regression to a degree.

In general, in simple regression our purpose is to minimize the error rate while SVR attempts to fit the error within a threshold. In order to better understand the working principle of SVR, the Figure 6 given below illustrates a one dimensional toy example along with introducing some concepts such as (a) boundary line (*the dotted line*), (b) hyperplane (*the separation line between data points belonging to different classes*) and (c) support vectors (*the data points constructing the boundary line*).

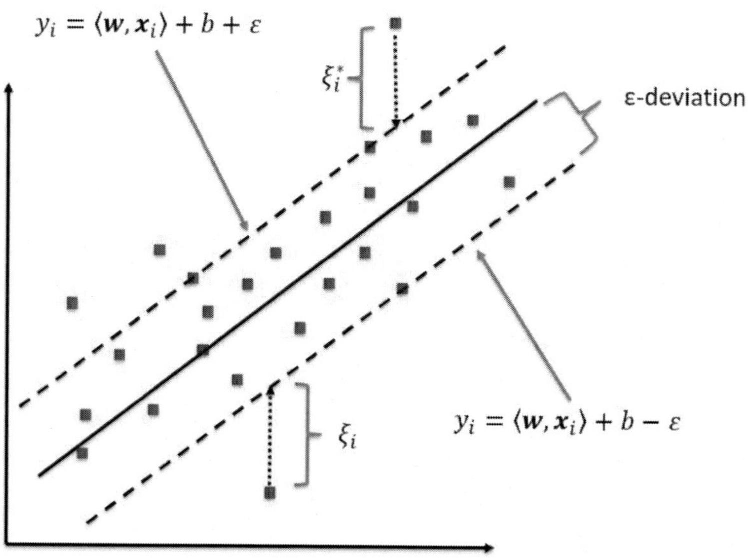

Figure 6. The illustration of one dimensional support vector regression function. Adopted from (Kleynhans et al., 2019).

If we assume the function of hyperplane as $\boldsymbol{W}x + b = 0$ then the equations of the boundary lines will be $\boldsymbol{W}x + b = +\varepsilon$ and $\boldsymbol{W}x + b = -\varepsilon$ respectively. Thus, the hyperplane equation becomes $\varepsilon \leq \boldsymbol{W}x + b \leq +\varepsilon$ along with the fact that $y = \boldsymbol{W}x + b$ and $y - \boldsymbol{W}x - b = 0$. Consequently, the ultimate purpose of the SVR optimization is to decide a decision boundary having e distance from the original hyperplane such that all the data points locatse at the closest hyperplane or support vectors bounded by boundary lines. If we investigate the Figure 6 above, we can see that the line shows the ground truth values whereas the points depict the predicted values of y.

Moreover, dashed lines constitute the bounds which are e distance away from the ground truth data. Support vector regression considers the points outside of the 'tube' (dashed lines) in order to create a model by solving (7) given below (Kleynhans et al., 2019):

$$\text{minimize} = \frac{1}{2}\|\mathbf{w}\|^2 + C\sum_{i=1}^{n}(\varphi_i^* + \varphi_i) \tag{7}$$

$$\text{subject to} \begin{cases} y_i - \langle \boldsymbol{w}, \boldsymbol{x_i} \rangle - b \leq \varepsilon + \varphi_i^* \\ \langle \boldsymbol{w}, \boldsymbol{x_i} \rangle + b - y_i \leq \varepsilon + \varphi_i \end{cases}$$

where $\boldsymbol{w}$ is the learned weight vector, $\boldsymbol{x}_i$ is the i-th training instance, y_i is the training value, φ_i is the distance between the bounds and predicted values outside the bounds (Kleynhans et al., 2019). The aim of C value here is the penalty controlling parameter that is set by user and helps the algorithm to avoid model overfitting by assigning a penalty score for the predicted values lie outside the bounds.

If the input data is not inherently linear separable then the use of kernel trick comes into prominence. By definition, every dot/inner product $\langle \boldsymbol{w}, \boldsymbol{x_i} \rangle$ is replaced with a radial basis function (RBF) kernel in order to project the input data into a higher dimensional space which holds the properties of the Hilbert spaces (Kleynhans et al., 2019).

Random Forests

Developed by Breiman (2001), Random Forests or Random Decision Forests are an ensemble learning method used for classification and regression tasks. The main advantage of random forests is avoiding the problem of overfitting to training data in conventional decision trees by constructing a large set of decision trees (e.g., 10- 1000) via random selection of input features and samples. In other words, it acts as a meta estimator which fits many decision trees on several sub-samples in order to enhance the prediction power either or classification or regression along with reducing the overfitting. Note that, the number of cases in each sub-sample is always the same as the input dataset by applying replacement.

The aforementioned strategy is called as *bagging* or *bootstrap aggregation* which can be summarized as follows. Suppose that we are given a training dataset $X = x_1, x_2, \dots, x_n$ that has class labels/real valued scores $Y = y_1, y_2, \dots, y_n$. At this point, the bagging chooses random samples, with replacement, m training set (i.e., X_m, Y_m) from original training data and construct classification or regression tree T_m for each of them. Following to training of each classifier or regressor, either majority voting of the predicted labels or taking the average of continuous scores yields the final outcome of the algorithm. The case for regression problems has been given in the expression (8) below:

$$Outcome(T) = \frac{1}{M}\sum_{m=1}^{M} T_m(x') \qquad (8)$$

It should be noted that, bootstrapping strategy improves the performance of the model since it reduces the variance along with avoiding increase of the bias. This is also an expected outcome due to having more than one tree which is sensitive to the noise in input data. When dealing with several trees, averaging the outputs yields a better estimation in case when these trees are not correlated.

XGBoost – Regularized Gradient Boosting

Initially proposed by Chen and Guestrin (2016), XGBoost is a highly optimized open source distributed gradient tree boosting framework which is designed in order to be efficient, portable and flexible on supervised learning problems (XGBoost, 2019). The name XGBoost refers to "Extreme Gradient Boosting" where the term "Gradient Boosting" comes from the seminal work of Friedman (2001) on "Greedy function approximation: A Gradient Boosting Machine" whereas the "Extreme" word stands for pushing the computational limits and optimizing the resource usage.

One key advantage of XGBoost is that it has been ported to various operating systems such as Windows™, Linux and macOS™. Further, it supports several distributed processing environments such as Hadoop and Apache Spark and is implemented for several programming languages such as Java, Python, R. After having data scientists achieve state of art results in many Kaggle competitions, the XGBoost package and its underlying gradient boosting method have gained more attention in order to obtain higher accuracy and lower bias as well.

XGBoost framework involves a parallel tree boosting algorithm (i.e., gradient boosted trees) which is universally applicable to various data science problems such as classification, regression, feature importance computation (Zheng, Yuang and Chen, 2017) and ranking (XGBoost, 2019). Similar to random forests method mentioned above, XGBoost uses tree ensembles which are consist of either classification or regression trees. As can be predicted easily, random forests and boosted trees share the same models. However, their key differences can be listed as follows: (a) the strategy on how they are trained, (b) a novel sparsity-aware algorithm for parallel tree learning, (c) weighted quantile sketch approach for proposing candidate tree split points by handling instance weights and finally (d) a cache-aware block structure for out-of-core tree learning (Chen and Guestrin, 2016). Moreover, techniques like data compression and sharding were also incorporated in order to create a scalable tree boosting solution.

In most simplified form, XGBoost employs objective functions which include measurement of the training loss and regularization term expressed in (9) and (10) where L denotes the training loss function and δ stands for the regularization term.

$$obj(\theta) = L(\theta) + \delta(\theta) \tag{9}$$

$$obj = \sum_{i=1}^{n} l\left(y_i, y_i^{*(t)}\right) + \sum_i^t \delta(f_i) \tag{10}$$

The regularization term is added in order to avoid overfitting on training data that yields a simple and predictive model (XGBoost Intro, 2019). If we suppose that the selected loss function is the mean square error (MSE) loss, then the term L turns into (11) that is given below:

$$L(\theta) = \sum_i (y_i - y_i^*)^2 \tag{11}$$

The MSE loss is a second order quadratic and inherently differentiable function so the gradient of the loss function can be computed as given in expression (12) presented below:

$$obj^{(t)} = \sum_{i=1}^{n} \left[2\left(y_i^{*(t-1)} - y_i\right) f_t(x_i) + f_t(x_i)^2\right] + \delta(f_t) + \text{constant} \tag{12}$$

In case of not easily differentiable loss functions (e.g., logistic loss), Taylor expansion of the loss function is computed. However, it is noteworthy that those kind of functions cost more in terms of computational time. As a summary, taking the derivatives of the loss functions for each tree generation yields a simple, but yet effective solution to find out the parameters of the trees. Nonetheless, as pointed out in (XGBoost Intro, 2019) learning a tree structure is much more complex compared to conventional optimization problems where only taking the gradient is enough. Therefore, creators of the XGBoost have employed "additive training" strategy in order deal with this problem. Thus, they are able to add a new tree based on the knowledge gained from previously created trees. Due to space constraints in this paper, for further reading on XGBoost framework, it is highly suggested to read the paper of Chen and Guestrin (2016).

Evaluation Metrics

Regarding the nature of the problem, throughout the experiments, we have employed the metrics of mean absolute error (MAE), mean square error (MSE), root mean square error (RMSE), coefficient of determination (R^2) and accuracy. If we denote the ground truth value as y_i, the predicted score as y_i^*, the count of samples under test as n and the average of the observed data $\bar{y}$, then the formulation of these metrics were given below in respective order. Note that, except the coefficient of determination and accuracy, better the machine learning models fit (i.e., generalize), lower the scores we have. On the other hand, metrics of accuracy and R^2 gets the value of 1 in ideal case.

$$\text{MAE} = \frac{1}{n}\sum_{i=1}^{n}|y_i - y_i^*| \qquad (13)$$

$$\text{MSE} = \frac{1}{n}\sum_{i=1}^{n}(y_i - y_i^*)^2 \qquad (14)$$

$$\text{RMSE} = \sqrt{\frac{1}{n}\sum_{i=1}^{n}\left(y_i - y_i^*\right)^2} \qquad (15)$$

$$R^2 = 1 - \frac{\sum_i(y_i - f_i)^2}{\sum_i(y_i - \bar{y})^2} \qquad (16)$$

$$\text{Accuracy} = 100 - \frac{1}{n}\left(100 \times \sum_{i=1}^{n}\left|\frac{y_i - y_i^*}{y_i}\right|\right) \qquad (17)$$

EXPERIMENTAL STUDIES

Data Preparation

In this study, the GSI values defined for eleven rock types at twenty outcrops were implemented (Figure 7). The data of lava and peridotite was obtained from Marinos et al., (2006), while sandstone-siltstone and siltstone-sandstone data was acquired from Marinos (2019). The rest of the data was provided by the General Directorate of Turkish State Railways of Turkey. The GSI values were determined according to the GSI schemes suggested by Hoek et al., (1998) and Marinos and Hoek (2001) (see Figure 1 and 2). Furthermore, the GSI data which was acquired from different data sources were defined between reasonable ranges, as suggested by Hoek et al., (1998) and Marinos and Hoek (2001). The meaningful GSI values that can be retrieved from these schemes are only multiples of five. For this reason, as can be seen from Figure 8, the GSI values, which were implemented in training and test stages were assigned as multiples of five. The number of the raw images having unique GSI values is 62.

However, although our corpus covers almost the entire range of GSI scope collected from different kinds of rocks, the number of total images (i.e., 62) was not sufficient to create a generalizable and robust machine learning model. For this reason, we have employed a technique called as "data augmentation" in order to increase the count of image sample. In this context, by using a single image, we first cropped four equal sized sub-images each originally located from top-left, top-right, bottom-left and bottom right regions of the input image. Moreover, we also acquired a center crop having the same width and height. Thus, at first stage, we have obtained six images from one single image. Next, we have increased the number of image samples by rotating each of the image by degrees of 90, 180 and 270. Consequently, we have increased the sample count by factor of 24 (i.e., 6×4) yielding the total number of image samples reaches up to 1488 (i.e., 62 × 24). Note that, data augmentation is a well-known and widely used technique in computer vision tasks since it reduces the bias and overfitting along with creating more robust models against natural rotations and transformations.

For the next step, we have split the overall dataset into a training and testing set covering 80% and 20% of the initial corpus respectively. As a result, our dataset has been formed such that 1193 training and 295 testing images exist.

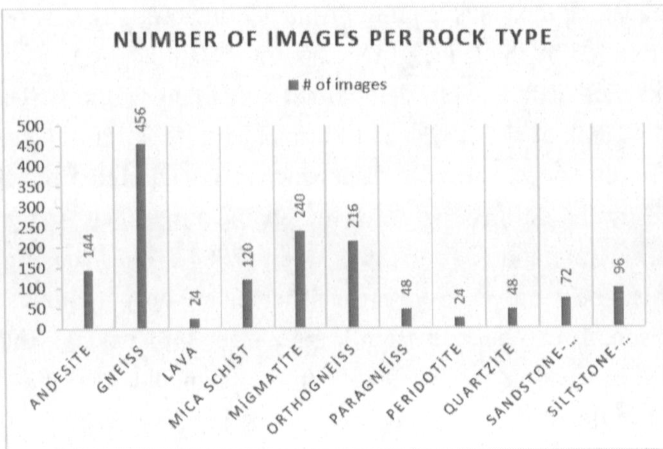

Figure 7. Distribution of rock types in our corpus.

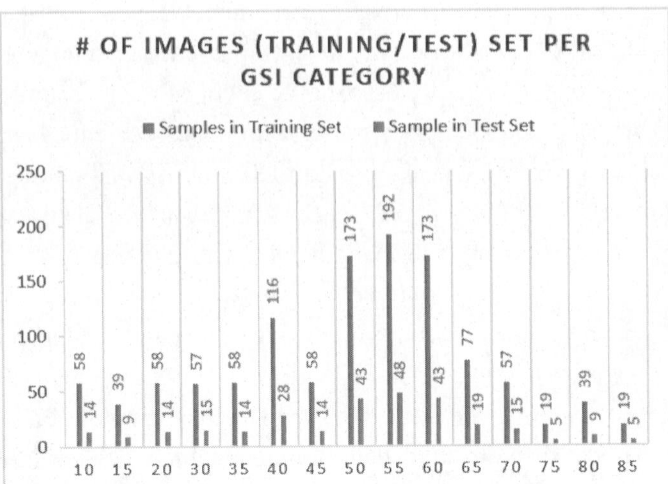

Figure 8. Distribution of rock images in training and test set per GSI categories.

Figure 9. Sample images from rock types. The images belonging to "Lava" and "Peridotite" were taken from Marinos et al., (2006) whereas "Sandstone-Siltstone" and "Siltstone-Sandstone" were collected from the work of Marinos (2019); the rest of the images were provided by the General Directorate of Turkish State Railways of Turkey.

Image Descriptor Generation

Image descriptors are generally classified into two groups whether they are local or global. Global image descriptors aim to extract a holistic and discriminative visual signature by considering whole image content while the local ones focus on extracting discriminative descriptions from important keypoints. In this work we have decided to use and experiment two well known local image descriptors namely SIFT and DAISY which the details of them given above. The main reasons of this preference is based on the nature of the problem.

First of all, we considered the need of scale invariance property of selected descriptors since the process of GSI value estimation from natural images requires extracting features from photographs having taken with ranging distances from the rock surface. Having robustness to geometric distortions and varying illumination conditions has also played key role to this selection. As stated before, SIFT has one key advantage over DAISY descriptor since it support rotational invariance. Thus, regarding the nature of problem, we have aimed to experiment whether rotational invariance property is important or not.

During the feature vector generation we have followed the bag-of-visual-words approach. For this purpose we have created a script in Python programming language which do the following operations listed below:

- Open each image in the training set
- Compute the key points and accumulate descriptors extracted around those key points
- Do a K-means clustering with the *k* parameter (i.e., *visual word count*) provided by user and obtain cluster centroids
- Following to computation of cluster centroids, assign each key point descriptor to the closest cluster centroid and increment related bin count by 1 yielding a frequency histogram having *k* bin for each image in training set
- Normalize feature vector of each image
- Having computed cluster centroids by utilizing training dataset, compute normalized frequency histograms for test images

- Output training and test feature vectors with ground truth GSI values to a comma separated value file for further machine learning based tasks.

The number of visual words is important in BOVW approach since it directly plays a key role on degree of sparsity of feature vectors and the discriminative power. At this point, we have followed the scheme described by Lazebnik et al., (2006). Accordingly, we have experimented various visual word counts such as k =50, 100, 200 and 400. We aimed, by following to this scheme, we have both revealed weak and strong features.

For SIFT based descriptor extraction, we used OpenCV Python module and run the algorithm with default parameters. On the other hand, for DAISY descriptors, an open-source framework namely "skimage" has been employed. Nonetheless, as stated before in section 3.1.2, DAISY descriptor is sensitive to its parameters and it has intended to be used with dense sampling strategy rather than sparse sampling. However, the optimum parameter set of DAISY for our problem domain was unknown. For this reason, we have primarily experimented the predictive performance of the method in order to understand whether we should sample the keypoints in dense or sparse manner. Our initial tests our training and test sets by using different *radius* (R) and *step size* (i.e., angular quantization no) values show that the prediction performance degrades dramatically as the sparsity of sampling increases. Indeed, this is a trade-off between the computation time and memory usage versus the accuracy obtained since dense sampling produces much more descriptors from input image. Therefore, we have chosen the dense sampling regime in favor of accuracy. As a result, we have set the parameters of the DAISY algorithm as follows: (a) radius = 8, (b) step_size = 16, (c) rings = 3, (d) histograms = 6 and (e) orientation = 8. According to this configuration, the algorithm produces 152 dimensional feature vectors on a virtually created grid where having 16 pixel wide cells and each crossing point is considered as a key point.

Prediction Models

Following to generation of image descriptors and transforming them to visual signatures of each image by using BOVW method, we have built various machine learning based regression models generated by support vector regression (SVR), random forests (RF) and XGBoost based regularized trees (XGB). As noted before, our traning and test sets contain 1193 and 295 natural rock images respectively. In order to build such machine learning models we have created a Python script based on open source "sklearn" libraries. At first stage, we have created these three kinds of models and evaluated their generalization performance on the training set itself.

During the building stage of SVR model, we have experimented both linear and radial basis kernels and set the cost (C) parameter as 100 whereas epsilon parameter as 0.2. On the other hand, for RF models we have not set any model specific hyperparameter thus enabling the algorithm to work with default parameters. In contrast, XGB requires several parameters such as maximum depth of trees and learning rate at startup. Therefore, we have set max_depth = 6, gamma = 0.4, learning rate = 0.08, number of estimators = 1000 and evaluation metric as mean absolute error (MAE). With these configurations, we have totally built 32 models involving two type of features (i.e., SIFT and DAISY), four different visual word count (i.e., 50, 100, 200, 400) and four different algorithm configuration (i.e., SVR-linear, SVR-rbf, RF and XGB). The results were given in Table 1 and Table 2 according the descriptor type we employed. Furthermore, we have listed the performance of the models in various metrics covering MAE, MSE, RMSE, R^2 and accuracy.

According to the results, XGBoost performs the best generalization (i.e., accuracy of 99.65%) for both of the descriptors. Moreover, except random forests, the obtained errors rates show decrease with the increase in number of visual words for the other algorithms. One another key finding is that the SIFT based features perform much better than DAISY descriptors. This results clearly shows that rotational invariance plays a key role for our domain. However, as described before, though they inspire from similar ideas, SIFT and DAISY descriptors work differently especially in terms of key point sampling and selection of region of interest (ROI). While SIFT analyzes 16×16 pixel regions as the context, the DAISY serves more options to define this ROI. Our sampling strategy covers dense sampling with covering more context (i.e., around 24 pixels). Therefore one possible reason of these results might be perhaps because of covering more context around key points.

One another important finding we have discovered is that, 'rbf' kernel is more suitable than the 'linear' kernel. This implies that the underlying algorithm requires a non linear optimization for solving the problem. Moreover, we also explored that XGB and RF handle the sparsity of the features more successfully. However, the boosting technique and gradient based learning yield a better performance for XGB instead of RF.

Table 1. Performance of SIFT descriptors with various visual words and machine learning methods on training images

Word Count	Method	MAE	MSE	RMSE	R^2	Acc.
50 words	SVR-linear	10.48	191.8	13.85	0.384	62.86%
100 words	SVR-linear	9.55	179.9	13.41	0.423	65.75%
200 words	SVR-linear	8.56	156.5	12.50	0.498	70.35%
400 words	SVR-linear	6.30	100.2	10.01	0.678	80.95%
50 words	SVR-rbf	7.02	103.1	10.15	0.669	78.65%
100 words	SVR-rbf	6.09	87.58	9.35	0.719	81.44%
200 words	SVR-rbf	4.92	63.26	7.95	0.797	85.80%
400 words	SVR-rbf	3.47	40.09	6.33	0.871	90.30%
50 words	Random F.	1.77	6.31	2.51	0.979	94.55%
100 words	Random F.	1.90	6.46	2.54	0.979	94.11%
200 words	Random F.	2.07	7.54	2.74	0.975	94.43%
400 words	Random F.	2.39	9.49	3.01	0.969	92.44%
50 words	XGBoost	0.16	0.03	0.19	0.999	99.58%
100 words	XGBoost	0.15	0.03	0.18	0.999	99.61%
200 words	XGBoost	0.14	0.03	0.17	0.999	99.63%
400 words	XGBoost	0.13	002	0.16	0.999	99.65%

Table 2. Performance of DAISY descriptors with various visual words and machine learning methods on training images

Word Count	Method	MAE	MSE	RMSE	R2	Acc.
50 words	SVR-linear	11.43	226.3	15.04	0.274	57.14%
100 words	SVR-linear	10.76	204.5	14.29	0.344	60.53%
200 words	SVR-linear	10.06	195.7	13.99	0.372	63.77%
400 words	SVR-linear	7.98	151.9	12.32	0.512	73.07%
50 words	SVR-rbf	6.71	103.3	10.16	0.668	78.69%
100 words	SVR-rbf	5.56	77.15	8.78	0.752	82.07%
200 words	SVR-rbf	4.11	54.50	7.38	0.825	85.86%
400 words	SVR-rbf	3.27	48.49	6.96	0.844	88.39%
50 words	Random F.	3.18	16.87	4.10	0.945	89.41%
100 words	Random F.	3.27	18.44	4.29	0.940	89.05%
200 words	Random F.	3.38	18.29	4.27	0.941	88.97%
400 words	Random F.	3.52	20.30	4.50	0.934	87.83%
50 words	XGBoost	0.14	0.03	0.18	0.999	99.60%
100 words	XGBoost	0.14	0.03	0.17	0.999	99.62%
200 words	XGBoost	0.13	0.02	0.17	0.999	99.62%
400 words	XGBoost	0.12	0.02	0.16	0.999	99.65%

Validation

In this stage, we have evaluated the performance of the models created with training images on test set. The results were given in Table 3 and Table 4 with bold marking at best models. Experiments were carried out on a computer having Intel 8750 processor with 24 GB of memory. The findings we have discovered from the results were listed below:

- The best performing machine learning method has been found as XGBoost
- The best performing descriptor – word size - model configuration has been found as SIFT, 50 words and XGBoost. Accordingly, we achieved to create a model performing 3.55 (MAE), 0.92 (R2) and accuracy of 90.33%.

Table 3. Performance of SIFT descriptors with various visual words and machine learning methods on test images. The best configuration is marked with bold style

Word Count	Method	MAE	MSE	RMSE	R2	Acc.
50 words	SVR-linear	10.84	204.05	14.28	0.339	60.65%
100 words	SVR-linear	10.18	188.53	13.73	0.389	65.32%
200 words	SVR-linear	10.77	193.35	14.66	0.303	63.45%
400 words	SVR-linear	11.21	236.75	15.38	0.233	64.74%
50 words	SVR-rbf	7.39	109.97	10.47	0.644	77.85%
100 words	SVR-rbf	6.71	94.95	9.74	0.692	79.85%
200 words	SVR-rbf	6.07	78.75	8.87	0.744	82.42%
400 words	SVR-rbf	5.21	53.79	7.33	0.825	85.14%
50 words	Random F.	4.56	41.26	6.42	0.866	86.32%
100 words	Random F.	4.69	37.99	6.16	0.876	86.44%
200 words	Random F.	5.25	44.01	6.63	0.857	84.02%
400 words	Random F.	6.01	58.61	7.65	0.810	81.02%
50 words	XGBoost	3.55	23.93	4.89	0.920	90.33%
100 words	XGBoost	4	29.08	5.39	0.905	88.67%
200 words	XGBoost	4.08	29.76	5.45	0.903	88.62%
400 words	XGBoost	4.7	37.21	6.1	0.879	85.71%

- The order success in employed estimators has been observed as XGBoost, RF and SVR in respective order
- SIFT features have clearly outperformed DAISY based features
- Similar to the results obtained on training data, 'rbf kernel performed better than 'linear' one in terms of all metrics
- Reduction of visual words have yielded better results for XGB and RF models while the vice versa happened for SVR based models. Combined with the superior success of XGB and RF over SVR, this result highlights the importance of model complexity and overfitting problem with limited sparse data. Likewise, XGB has performed better with strong features while SVR has required more weak features to increase the generalization capability. This finding is also in favor of XGB since working with less visual

words costs much less in terms of computation resource which enables us to build a more efficient and effective solution

- Including the steps of feature generation, normalization and prediction, the overall prediction time for an image was recorded as 0.34 sec. in average

The results belonging to test set clearly reveal that SIFT descriptors combined with the estimator of XGBoost can be used efficiently and effectively in our problem domain. Moreover, the average prediction time for a single image also enables us to use the model in realtime scenarios.

Table 4. Performance of DAISY descriptors with various visual words and machine learning methods on test images. The best configuration is marked with bold style

Word Count	Method	MAE	MSE	RMSE	R2	Acc.
50 words	SVR-linear	12.00	238.11	15.43	0.228	57.91%
100 words	SVR-linear	12.72	264.76	16.27	0.142	56.24%
200 words	SVR-linear	13.24	274.22	16.55	0.111	57.42%
400 words	SVR-linear	15.00	355.05	18.84	0.093	53.77%
50 words	SVR-rbf	8.59	124.9	11.17	0.595	73.68%
100 words	SVR-rbf	8.52	116.4	10.78	0.622	73.69%
200 words	SVR-rbf	8.24	112.88	10.63	0.634	73.99%
400 words	SVR-rbf	9.67	146.88	12.11	0.524	69.92%
50 words	Random F.	8.66	122.29	11.05	0.585	72.17%
100 words	Random F.	8.92	131.19	11.45	0.575	70.39%
200 words	Random F.	8.95	135.22	11.62	0.562	70.22%
400 words	Random F.	9.36	136.22	11.67	0.558	68.77%
50 words	XGBoost	7.85	96.83	9.84	0.686	77.08%
100 words	XGBoost	8.36	115.25	10.73	0.626	74.24%
200 words	XGBoost	8.79	124.27	11.14	0.597	71.36%
400 words	XGBoost	9.24	137.46	11.72	0.554	70.10%

Discussion and Conclusion

We attempted to develop a system which is able to calculate approximate GSI value of the rock mass when we only give the photograph of the rock outcrop to the algorithm. We, in fact, expected from the system to behave an experienced engineering geologist. The results obtained in the experimental studies are promising. The accuracy values were obtained as higher than 99% and 90% for training and test images respectively, by using the SIFT descriptor and XGBoost estimator. Previous investigations were also conducted to indirectly estimate GSI values (Sonmez et al., 2003; Poulsen et al., 2015; Hong et al., 2017). However, in these methods the parameters need to be either calculated or defined in advance. Unfortunately, additional requirements for calculations and definitions prevent the method from being automated and hamper the human and computer interaction.

In this sense, the method proposed in this study is fully automated. Eleven rock types at twenty outcrops were implemented in this study. As the number of rock types evaluated in the corpus and the number of GSI definitions evaluated at the training stage increase, the sensitivity of the proposed method, in other words, the capacity to generalize will obviously increase.

When we compare the observed GSI values with the values predicted by using SIFT descriptor and XGBoost estimator on the test images, it is revealed the predicted and observed value differences concentrated between –5 and +5 lines (Figure 10). This result indicates that the proposed method is quite acceptable. However, there are still some outliers; e.g., five estimations have a difference value of higher than –10. The main reasons for this peculiarity raised from the quality of the photos and the class imbalance problem lying in the dataset. If we investigate the outliers plotted in Figure 10 and the distribution of image samples across different GSI values (Figure 8), it can be easily seen that, the outlier points in results come from the GSI categories we have undersampled. In case where texture and patterns are not able to be retrieved from the image of the rock outcrop, image descriptors are not able to be generated properly. This obviously results in poor training and estimation.

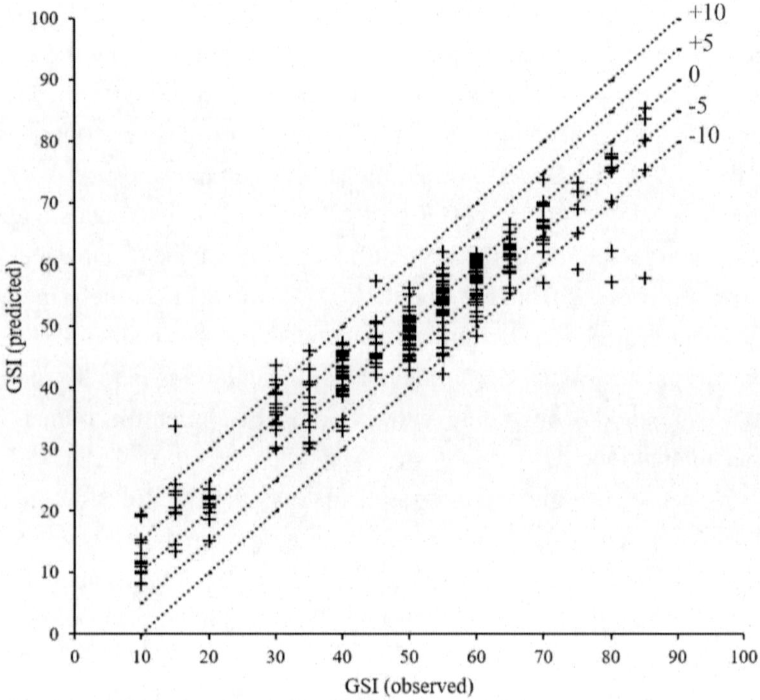

Figure 10. Comparison of the observed GSI values with the values predicted by using SIFT descriptor and XGBoost estimator on the test images; [predicted–observed] lines are also indicated on the graph.

The future studies will include increasing the number of rock type evaluated in the system, increasing the number of GSI determinations, and investigating different machine learning methods and image descriptors as well. Furthermore, the development of different estimation models for each rock type will significantly increase the estimation performance and generalization capacity of the system. Additionally, in a very recent paper, Hoek and Brown (2019) suggested that the range of GSI values should also be treated as a distribution. By considering this recommendation, the GSI scheme problem will also be handled as classification problem, and deep learning methods will be investigated.

REFERENCES

Afroz, S. and Rachel Greenstadt, "PhishZoo: Detecting Phishing Websites by Looking at Them," Paper presented at *IEEE Fifth International Conference on Semantic Computing*, Palo Alto, USA, 2011.

Barton, N. R., R. Lien, J. Lunde, "Engineering classification of rock masses for the design of tunnel support," *Rock Mechanics* 6, no 4, (1974).

Bieniawski, Z. T., "Engineering classification of jointed rock masses," *Trans. S. Afr. Inst. Civ. Eng* 15, (1973).

Breiman, Leo, "Random Forests", *Machine Learning* 45, no 1, (2001) 5–32. doi:10.1023/A:1010933404324.

Cai, M., P. K. Kaiser, H. Uno, Y. Tasaka, M. Minami, "Estimation of rock mass strength and deformation modulus of jointed hard rock masses using the GSI system" In*ternational Journal of Rock Mechanics and Mining Science*s 41, no 1, 2004.

Chen, Tianqi and Carlos Guestrin, *"XGBoost: A Scalable Tree Boosting System",* Paper presented at KDD'16, San Francisco, USA, 2016.

Deere, D. U., "Technical description of rock cores for engineering purposes," *Rock Mechanics and Rock Engineering* 1, no 1, (1964).

Friedman, Jerome, H. "Greedy Function Approximation: A Gradient Boosting Machine," *The Annals of Statistics* 29, no 5, (2001).

Foo, Jun. J. and Ranjan Sinha, "Pruning sift for scalable near-duplicate image matching," Paper presented at Eighteenth Conference on Austrialasian Database Conference, Ballarat, Austraila, 2007.

Fuiorea, Daniela et al., "Sensor Node Localization using SIFT Algorithm," Paper presented at *9th WSEAS International Conference on Automation and Information,* Bucharest, Romania, 2008.

Guo, Jing-Ming, Yun-Fu, Liu, Zong-Jhu, We, "Duplication forgery detection using improved DAISY descriptor," *Expert Systems with Applications* 40, no 2 (2013). doi: 10.1016/j.eswa.2012.08.002.

Hoek, E., "Putting numbers to geology—an engineer's view point," *Quarterly Journal of Engineering Geology* 32, (1999).

Hoek, E. and E. T. Brown, "Underground Excavations in Rock," *The Institution of Mining Metallurgy*, London, (1980).

Hoek, E. and E. T. Brown, "The Hoek–Brown failure criterion—a 1988 update", Paper presented at *1^5th^ Canadian Rock Mechanics Symposium*, University of Toronto, 1988.

Hoek, E. and E. T. Brown, "Practical estimates of rock mass strength," *International Journal of Rock Mechanics and Mining Sciences & Geomechanics Abstracts* 34, pp.1165–1186, 1997.

Hoek, E. and E. T. Brown, "The Hoek–Brown failure criterion and GSI – 2018 edition," *Journal of Rock Mechanics and Geotechnical Engineering* 11, (2019).

Hoek, E., T. G. Carter, M. S. Diederichs, "Quantification of the Geological Strength Index Chart", Paper presented *at 47th US Rock Mechanics/Geomechanics Symposium*, San Francisco, CA, USA, 2013.

Hoek, E., P. Marinos, M. Benissi, "Applicability of the geological strength index (GSI) classification for very weak and sheared rock masses: the case of the Athens schist formation," *Bulletin of Engineering Geology and the Environment* 57, (1998).

Hong, K., E. Han, K. Kang, "Determination of geological strength index of jointed rock mass based on image processing," *Journal of Rock Mechanics and Geotechnical Engineering* 6, (2017).

Karami, E., et al., "Image Identification Using SIFT Algorithm: Performance Analysis against Different Image Deformations," Paper presented at *2015 Newfoundland Electrical and Computer Engineering Conference,* St.John's, Canada, 2015.

Keser, Reyhan. K. et al., "Vehicle Logo Recognition with Reduced-Dimension SIFT Vectors Using Autoencoders," Paper presented at International Workshop on Computational Intelligence for Multimedia Understanding, Kos Island, Greece, 2017.

Kleynhans, Tania, Matthew, Montanaro, Aaron Gerace and Christopher Kanan. Predicting Top-of-Atmosphere ThermalRadiance Using MERRA-2 Atmospheric Datawith Deep Learning, 2017. Accessed June 11, 2019. https://www.researchgate.net/publication/320916953_

Predicting_Top-of-Atmosphere_Thermal_Radiance_Using_MERRA-2_Atmospheric_Data_with_Deep_Learning.

Lazebnik, Svetlana, Cordeila Schmidt and Jean Ponce, "Beyond Bags of Features: Spatial Pyramid Matching for Recognizing Natural Scene Categories," Paper presented at *CVPR 2006*, Newyork, USA, 2006.

Li, Li and John Zic. "Image Matching Algorithm based on Feature-point and DAISY Descriptor", *Journal of Multimedia* 9, no. 6 (2014). doi: 10.4304/jmm.9.6.829-834.

Lowe, David G. "Distinctive image features from scale invariant keypoints," *International Journal of Computer Vision* 60, no. 2 (2004). doi:10.1023/B:VISI.0000029664.99615.94.

Marinos, V., P. Marinos, E. Hoek, "The geological strength index: applications and limitations", *Bulletin of Engineering Geology and the Environment* 64, (2005).

Marinos P., E. Hoek, "Estimating the geotechnical properties of heterogeneous rock masses such as flysch," *Bulletin of Engineering Geology and the Environment* 60, (2001).

OpenCV. "Introduction to SIFT (Scale-Invariant Feature Transform)" Last modified May 25, 2013. https://opencv-python-tutroals.readthedocs.io/en/latest/py_tutorials/py_feature2d/py_sift_intro/py_sift_intro.html.

Poulsen, B. A., D. P. Adhikary, M. Elmouttie, K. A. Wilkins, "Convergence of synthetic rock mass modeling and the Hoek-Brown strength criterion," *International Journal of Rock Mechanics and Mining Sciences* 80, (2015).

Russo, G., "Improving the reliability of GSI estimation: the integrated GSI-RMi system", Paper presented at ISRM Workshop Underground Works under Special Conditions, Madrid, 2007.

Russo, G., "A new rational method for calculating the GSI," *Tunnelling and Underground Space Techno*logy 24, (2009).

Sachdeva, Varsha D. et al., "Performance Evaluation of SIFT and Convolutional Neural Network for Image Retrieval," *International Journal of Advanced Computer Science and Applications* 8, no. 12 (2017). doi: 10.14569/IJACSA.2017.081268.

Sonmez, H., C. Gokceoglu, R. Ulusay, "An application of fuzzy sets to the Geological Strength Index (GSI) system used in rock engineering," *Engineering Applications of Artificial Intelligence* 16, (2003).

Sonmez, H. and R. Ulusay, "Modifications to the geological strength index (GSI) and their applicability to the stability of slopes," *International Journal of Rock Mechanics and Mining Sciences* 36, (1999).

Tola, Engin, Vincent Lepetit and Pascal Fua, "DAISY: An Efficient Dense Descriptor Applied to Wide-Baseling Stereo," *IEEE Transactions on Pattern Analysis and Machine Learning* 32, no. 5 (2010). doi: 10.1109/TPAMI.2009.77.

Tola Engin, "DAISY: A Fast Descriptor for Dense Wide Baseline Stereo and Multiview Reconstruction" PhD diss., Ecole Polytechnique Federale de Lausanne, Santa Barbara, 2010, http://www.engintola.com/research/thesis_tola_phd.pdf.

Vapnik, V. "The natura of Statistcial Learning Theory," Springer, New York, 1995.

Yang, Hongsheng, Wen-Yan Lin and Jiangbo Lu, "DAISY Filter Flow: A Generalized Discrete Approach to Dense Coresspondances," Paper presented at *CVP*R, Columbus, USA, 2014.

XGBoost Develoopers, "XGBoost Documentation," May 26, 2019. https://xgboost.readthedocs.io/en/latest/.

XGBoost Intro, "Introduction to Boosted Trees", June 9, 2019. https://xgboost.readthedocs.io/en/latest/tutorials/model.html.

Zhang, L., *"Engineering Properties of Rocks,"* In: Hudson, J. A. (Ed.), Elsevier Geo-Engineering Book Series, vol. 4. Elsevier, Oxford, (2005).

Zheng, Huiting, Jiabin Yuan and Long Chen, "Short-Term Load Forecasting Using EMD-LSTM Neural Networks with a Xgboost Algorithm for Feature Importance Evaluation," Energies 10, no. 8, (2017) doi: 10.3390/en10081168.

Zhu, Chao, Charles-Edmond Bichot and Liming Chen, "Visual object recognition using DAISY descriptor," Paper presented at IEEE International Conference on Multimedia Expo, Barcelona, Spain, 2011.

AUTHOR'S CONTACT INFORMATION

Turgut Ozseven
Department of Computer Engineering
Tokat Gaziosmanpasa University, Tokat, Turkey
Email: turgutozseven@gmail.com

INDEX

F

G

H

I

J

K

L

M

S

T

U

V

W

Y